FAO
FISHERIES AND
AQUACULTURE
TECHNICAL
PAPER

531

Use of algae and aquatic macrophytes as feed in small-scale aquaculture

A review

by

Mohammad R. Hasan
Aquaculture Management and Conservation Service
Fisheries and Aquaculture Management Division
FAO Fisheries and Aquaculture Department
Rome, Italy

and

Rina Chakrabarti
University of Delhi
Delhi, India

FOOD AND AGRICULTURE ORGANIZATION OF THE UNITED NATIONS
Rome, 2009

The designations employed and the presentation of material in this information
product do not imply the expression of any opinion whatsoever on the part of the
Food and Agriculture Organization of the United Nations (FAO) concerning the
legal or development status of any country, territory, city or area or of its authorities,
or concerning the delimitation of its frontiers or boundaries. The mention of specific
companies or products of manufacturers, whether or not these have been patented, does
not imply that these have been endorsed or recommended by FAO in preference to
others of a similar nature that are not mentioned.

The views expressed in this information product are those of the author(s) and do not
necessarily reflect the views of FAO.

ISBN 978-92-5-106420-7

All rights reserved. Reproduction and dissemination of material in this information
product for educational or other non-commercial purposes are authorized without
any prior written permission from the copyright holders provided the source is fully
acknowledged. Reproduction of material in this information product for resale or other
commercial purposes is prohibited without written permission of the copyright holders.
Applications for such permission should be addressed to:
Chief
Electronic Publishing Policy and Support Branch
Communication Division
FAO
Viale delle Terme di Caracalla, 00153 Rome, Italy
or by e-mail to:
copyright@fao.org

© FAO 2009

Preparation of this document

Recognizing the increasing importance of the use of aquatic macrophytes as feed in small-scale aquaculture, the global review on this topic was undertaken as a part of the regular work programme of the Fisheries and Aquaculture Department of the Food and Agriculture Organization of the United Nations (FAO) by the Aquaculture Management and Conservation Service 'Study and analysis of feed and nutrients (including fertilizers) for sustainable aquaculture development'. This was carried out under the programme entity 'Monitoring, Management and Conservation of Resources for Aquaculture Development'.

The manuscript was edited for technical content by Michael B. New. For consistency and conformity, scientific and English common names of fish species were used from FishBase (www.fishbase.org/home.htm). Most of the photographs in the manuscripts were provided by the first author. Where this is not the case, due acknowledgements are made to the contributor(s) or the source(s).

Special thanks are due to Dr Albert G.J. Tacon (Universidad de Las Palmas de Gran Canaria, Spain), Dr M.A.B. Habib (Bangladesh Agricultural University, Bangladesh), Md. Ghulam Kibria (Ministry of Fisheries and Marine Resources, Namibia) and Dr Khondker Moniruzzaman (University of Dhaka, Bangladesh) who kindly provided papers and information. The Royal Netherlands Embassy in Dhaka, Bangladesh is acknowledged for kindly providing the reports of the Duckweed Research Project.

We acknowledge Ms Tina Farmer and Ms Françoise Schatto for their assistance in quality control and FAO house style and Mr Juan Carlos Trabucco for layout design. The publishing and distribution of the document were undertaken by FAO, Rome.

Mr Jiansan Jia, Service Chief, and Dr Rohana P. Subasinghe, Senior Fishery Resources Officer (Aquaculture), Aquaculture Management and Conservation Service of the FAO Fisheries and Aquaculture Department are also gratefully acknowledged for their support.

Abstract

This technical paper presents a global review on the use of aquatic macrophytes as feed for farmed fish, with particular reference to their current and potential use by small-scale farmers. The review is organized under four major divisions of aquatic macrophytes: algae, floating macrophytes, submerged macrophytes and emergent macrophytes. Under floating macrophytes, *Azolla*, duckweeds and water hyacinths are discussed separately; the remaining floating macrophytes are grouped together and are reviewed as 'other floating macrophytes'. The review covers aspects concerned with the production and/or cultivation techniques and use of the macrophytes in their fresh and/or processed state as feed for farmed fish. Efficiency of feeding is evaluated by presenting data on growth, food conversion and digestibility of target fish species. Results of laboratory and field trials and on-farm utilization of macrophytes by farmed fish species are presented. The paper provides information on the different processing methods employed (including composting and fermentation) and results obtained to date with different species throughout the world with particular reference to Asia. Finally, it gives information on the proximate and chemical composition of most commonly occurring macrophytes, their classification and their geographical distribution and environmental requirements.

Hasan, M.R.; Chakrabarti, R.
Use of algae and aquatic macrophytes as feed in small-scale aquaculture: a review.
FAO Fisheries and Aquaculture Technical Paper. No. 531. Rome, FAO. 2009. 123p.

Contents

Preparation of this document   iii
Abstract   iv
Abbreviations and acronyms   vii

Introduction   1

1. Algae   3
Classification   3
Characteristics   4
Production   7
Chemical composition   8
Use as aquafeed   8

2. Floating aquatic macrophytes – *Azolla*   17
Classification   17
Characteristics   17
Production   19
Chemical composition   21
Use as aquafeed   21

3. Floating aquatic macrophytes – duckweeds   29
Classification   29
Characteristics   31
Production   34
Chemical composition   40
Use as aquafeed   43

4. Floating aquatic macrophytes – water hyacinths   53
Classification   53
Characteristics   53
Production   54
Chemical composition   55
Use as aquafeed   55

5. Floating aquatic macrophytes – others   67
Classification   67
Characteristics   67
Production   68
Chemical composition   68
Use as aquafeed   70

6. Submerged aquatic macrophytes	**75**
Classification	75
Characteristics	75
Production	76
Chemical composition	76
Use as aquafeed	78
7. Emergent aquatic macrophytes	**89**
Classification	89
Characteristics	90
Production	91
Chemical composition	91
Use as aquafeed	93
8. Conclusions	**95**
Algae	95
Azolla	95
Duckweeds	96
Water hyacinths	96
Other floating macrophytes	97
Submerged macrophytes	98
Emergent macrophytes	99
References	**101**
Annex 1 Essential amino acid composition of aquatic macrophytes	119
Annex 2 Periphyton	123

Abbreviations and acronyms

APD	Apparent Protein Digestibility
BFRI	Bangladesh Fishery Research Institute
BW	Body Weight
DM	Dry Matter basis
DW	Dry weight
DWRP	Duckweed Research Project (Bangladesh)
EAA	Essential Amino Acid
FCR	Feed Conversion Ratio
FW	Fresh Weight
MAEP	Mymensingh Aquaculture Extension Project
MP	Muriate of Potash
NFE	Nitrogen-Free Extract
NGO	Non-governmental organization
PRISM	Project in Agriculture, Rural Industry Science and Medicine (an NGO)
SGR	Specific Growth Rate
TKN	Total Kjeldahl Nitrogen
TSP	Triple Super Phosphate
UASB	Upflow Anaerobic Sludge Blanket Reactor
2,4-D	2,4-Dichhlorophenoxyacetic acid

Introduction

Using feeds in aquaculture (sometimes referred to as aquafeeds) generally increases productivity. However, to maximize cost-effectiveness, it is particularly useful in small-scale aquaculture to utilize locally available materials, either as ingredients (raw materials) in compound aquafeeds or as sole feedstuffs.

There is also a vital need to seek effective ingredients that can either partially or totally replace marine ingredients as protein sources in animal feedstuffs generally, in particular in aquafeeds. While this broad topic is not dealt with in this review, many introductions to the literature of the past two decades are available, including New and Csavas (1995), Tacon (1994; 2004;), Tacon, Hasan and Subasinghe (2006), Tacon and Metain (2008), New and Wijkstrom (2002), FAO (2008) and Huntington and Hasan (2009).

This review deals with the characteristics of aquatic raw materials for use as feeds in small-scale aquaculture, namely algae (principally macro-algae – commonly referred to as seaweeds) and aquatic macrophytes. Aquatic macrophytes are aquatic plants that are large enough to be seen by the naked eye. They grow in or near water and are floating, submerged, or emergent.

Information includes current and potential usage of these materials by small-scale aquafarmers for target fish and crustaceans, together with details on their classification, characteristics (including such factors as their natural distribution and environmental requirements), production and chemical composition.

The review has been divided into seven major sections: one dealing with algae; four sections on floating macrophytes (namely *Azolla*, duckweeds, water hyacinths and others); a section on submerged macrophytes; and one on emergent macrophytes. Finally, the review contains a concluding section which summarizes previous chapters.

1. Algae

Algae have been used in animal and human diets since very early times. Filamentous algae are usually considered as 'macrophytes' since they often form floating masses that can be easily harvested, although many consist of microscopic, individual filaments of algal cells. Algae also form a component of periphyton, which not only provides natural food for fish and other aquatic animals but is actively promoted by fishers and aquaculturists as a means of increasing productivity. This topic is not dealt with in this section, since periphyton is not solely comprised of algae and certainly cannot be regarded as macroalgae. However, some ancillary information on this topic is provided in Annex 2 to assist with further reading. Marine 'seaweeds' are macro-algae that have defined and characteristic structures.

Microalgal biotechnology only really began to develop in the middle of the last century but it has numerous commercial applications. Algal products can be used to enhance the nutritional value of food and animal feed owing to their chemical composition; they play a crucial role in aquaculture. Macroscopic marine algae (seaweeds) for human consumption, especially *nori* (*Porphyra* spp.), *wakame* (*Undaria pinnatifida*), and *kombu* (*Laminaria japonica*), are widely cultivated algal crops. The most widespread application of microalgal culture has been in artificial food chains supporting the husbandry of marine animals, including finfish, crustaceans, and molluscs.

The culture of seaweed is a growing worldwide industry, producing 14.5 million tonnes (wet weight) worth US$7.54 billion in 2007 (FAO, 2009). The use of aquatic macrophytes in treating sewage effluents has also shown potential. In recent years, macroalgae have been increasingly used as animal fodder supplements and for the production of alginate, which is used as a binder in feeds for farm animals. Laboratory investigations have also been carried out to evaluate both algae and macroalgae as possible alternative protein sources for farmed fish because of their high protein content and productivity.

Microalgae and macroalgae are also used as components in polyculture systems and in remediation; although these topics are not covered in this paper, information on bioremediation is contained in many publications, including Msuya and Neori (2002), Zhou *et al.* (2006) and Marinho-Soriano (2007). Red seaweed (*Gracilaria* spp.) and green seaweed (*Ulva* spp.) have been found to suitable species for bioremediation. The use of algae in integrated aquaculture has also been recently reviewed by Turan (2009).

1.1 CLASSIFICATION

The classification of algae is complex and somewhat controversial, especially concerning the blue-green algae (Cyanobacteria), which are sometimes known as blue-green bacteria or Cyanophyta and sometimes included in the Chlorophyta. These topics are not covered in detail this document. However, the following provides a taxonomical outline of algae.

Archaeplastida
- Chlorophyta (green algae)
- Rhodophyta (red algae)
- Glaucophyta

Rhizaria, Excavata
- Chlorarachniophytes

- Euglenids

Chromista, Alveolata
- Heterokonts
  - Bacillariophyceae (diatoms)
  - Axodine
  - Bolidomonas
  - Eustigmatophyceae
  - Phaeophyceae (brown algae)
  - Chrysophyceae (golden algae)
  - Raphidophyceae
  - Synurophyceae
  - Xanthophyceae (yellow-green algae)
- Cryptophyta
- Dinoflagellates
- Haptophyta

The following sections discuss the characteristics and use of both 'true' algae and the Cyanophyta, hereinafter referred to as 'blue-green algae').

1.2 CHARACTERISTICS

Filamentous algae and seaweeds have an extremely wide panorama of environmental requirements, which vary according to species and location. Ecologically, algae are the most widespread of the photosynthetic plants, constituting the bulk of carbon assimilation through microscopic cells in marine and freshwater.

The environmental requirements of algae are not discussed in detail in this document. However, the most important parameters regulating algal growth are nutrient quantity and quality, light, pH, turbulence, salinity and temperature. Macronutrients (nitrate, phosphate and silicate) and micronutrients (various trace metals and the vitamins thiamine (B_1), cyanocobalamin (B_{12}) and biotin) are required for algal growth (Reddy et al., 2005). Light intensity plays an important role, but the requirements greatly vary with the depth and density of the algal culture. The pH range for most cultured algal species is between 7 and 9; the optimum range is 8.2–8.7. Marine phytoplankton are extremely tolerant to changes in salinity. In artificial culture, most grow best at a salinity that is lower than that of their native habitat. Salinities of 20–24 ppt are found to be optimal. Lapointe and Connell (1989) suggested that the growth of the green filamentous alga *Cladophora* was limited by both nitrogen and phosphorus, but the former was the primary factor. Hall and Payne (1997) found that another green filamentous alga, *Hydrodictyon reticulatum*, had a relatively low requirement for dissolved inorganic nitrogen in comparison with other species. Rafiqul, Jalal and Alam (2005) found that the optimum environment for *Spirulina platensis* under laboratory conditions was 32 °C, 2 500 lux and pH 9.0. Further information on the environmental requirements of algae cultured for use in aquaculture hatcheries is contained in Lavens and Sorgeloos (1996). The environmental requirements of cultured seaweeds are discussed by McHugh (2002, 2003).

A brief description of some of the filamentous algae and seaweeds that have been used for feeding fish, as listed in Tables 1.1–1.3, is provided in the following subsections.

1.2.1 Filamentous algae

Filamentous algae are commonly referred to as 'pond scum' or 'pond moss' and form greenish mats upon the water surface. These stringy, fast-growing algae can cover a pond with slimy, lime-green clumps or mats in a short period of time, usually beginning their growth along the edges or bottom of the pond and 'mushrooming' to the surface. Individual filaments are a series of cells joined end to end which give the thread-like appearance. They also form fur-like growths on submerged logs, rocks and even on animals. Some forms of filamentous algae are commonly referred to as 'frog spittle' or 'water net'.

Spirulina, which is a genus of cyanobacteria that is also considered to be a filamentous blue-green algae, is cultivated around the world and used as a human dietary supplement, as well as a whole food. It is also used as a feed supplement in the aquaculture, aquarium, and poultry industries (Figure 1.1).

FIGURE 1.1
Spirulina sp.

Source: scienceblogs.com/energy/2008/08/big_energy_fr...

FIGURE 1.2
Spirogyra sp.

Source: Wim van Egmont©

Spirogyra, one of the commonest green filamentous algae (Figure 1.2), is named because of the helical or spiral arrangement of the chloroplasts. There are more than 400 species of *Spirogyra* in the world. This genus is photosynthetic, with long bright grass-green filaments having spiral-shaped chloroplasts. It is bright green in the spring, when it is most abundant, but deteriorates to yellow. In nature, *Spirogyra* grows in running streams of cool freshwater, and secretes a coating of mucous that makes it feel slippery. This freshwater alga is found in shallow ponds, ditches and amongst vegetation at the edges of large lakes. Under favourable conditions, *Spirogyra* forms dense mats that float on or just beneath the surface of the water. Blooms cause a grassy odour and clog filters, especially at water treatment facilities.

Cladophora (Figure 1.3) is a green filamentous algae that is a member of the Ulvophyceae and is thus related to the sea lettuce (*Ulva* spp.). The genus *Cladophora* has one of the largest number of species within the macroscopic green algae and is also among the most difficult to classify taxonomically. This is mainly due to the great variations in appearance, which are significantly affected by habitat, age and environmental conditions. These algae, unlike *Spirogyra*, do not conjugate (form bridges between cells) but simply branch.

FIGURE 1.3
Cladophora sp.

Source: Biopix.dk©

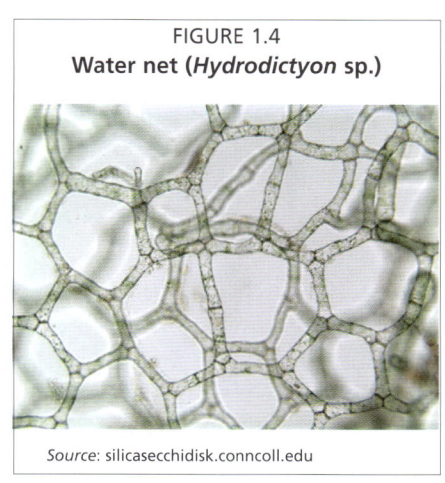

FIGURE 1.4
Water net (*Hydrodictyon* sp.)

Source: silicasecchidisk.conncoll.edu

Another green filamentous alga, *Hydrodictyon*, commonly known as 'water net', belongs to the family Hydrodictyaceae and prefers clean, eutrophic water. Its name refers to its shape, which looks like a netlike hollow sack (Figure 1.4) and can grow up to several decimetres.

1.2.2 Seaweeds

Ulva are thin flat green algae growing from a discoid holdfast that may reach 18 cm or more in length, though generally much less, and up to 30 cm across. The membrane is two cells thick, soft and translucent and grows attached (without a stipe) to rocks by a small disc-shaped holdfast. The water lettuce (*Ulva lactuca*) is green to dark green in colour (Figure 1.5). There are other species of *Ulva* that are similar and difficult to differentiate.

FIGURE 1.5
Sea lettuce (*Ulva lactuca*)

Source: Mandy Lindeberg (www.seaweedsofalaska.com)

FIGURE 1.6
Eucheuma cottonii

Source: www.actsinc.biz/seaweed.html

It is important to recognize that the genera *Eucheuma* and *Kappaphycus* are normally grouped together; their taxonomical classification is contentious. These are red seaweeds and are often very large macroalgae that grow rapidly. The systematics and taxonomy of *Kappaphycus* and *Eucheuma* (Figure 1.6) is confused and difficult, due to morphological plasticity, lack of adequate characters to identify species and the use of commercial names of convenience. These taxa are geographically widely dispersed through cultivation (Zuccarello *et al.*, 2006). These red seaweeds are widely cultivated, particularly to provide a source of carageenan, which is used in the manufacture of food, both for humans and other animals.

Gracilaria is another genus of red algae (Figure 1.7), most well-known for its economic importance as a source of agar, as well as its use as a food for humans.

FIGURE 1.7
Gracilaria sp.

Source: Eric Moody© (Wikipedia)

FIGURE 1.8
Porphyra tenera

Source: http://www.fao.org/fishery/species/search/en

The red seaweed *Porphyra* (Figure 1.8) is known by many local names, such as laver or *nori*, and there are about 100 species. This genus has been cultivated extensively in many Asian countries and is used to wrap the rice and fish that compose the Japanese food *sushi*, and the Korean food *gimbap*. It is also used to make the traditional Welsh delicacy, laverbread.

1.3 PRODUCTION

As in the case of their environmental conditions, the methods for culturing filamentous algae and seaweeds vary widely, according to species and location. This topic is not covered in this review but there are many publications available on algal culture generally, such as the FAO manual on the production of live food for aquaculture by Lavens and Sorgeloos (1996). Concerning seaweed culture, the following summary of the techniques used has been has been extracted from another FAO publication (McHugh, 2003) and further reading on seaweed culture can also be found in McHugh (2002).

> Some seaweeds can be cultivated vegetatively, others only by going through a separate reproductive cycle, involving alternation of generations.
>
> In vegetative cultivation, small pieces of seaweed are taken and placed in an environment that will sustain their growth. When they have grown to a suitable size they are harvested, either by removing the entire plant or by removing most of it but leaving a small piece that will grow again. When the whole plant is removed, small pieces are cut from it and used as seedstock for further cultivation. The suitable environment varies among species, but must meet requirements for salinity of the water, nutrients, water movement, water temperature and light. The seaweed can be held in this environment in several ways: pieces of seaweed may be tied to long ropes suspended in the water between wooden stakes, or tied to ropes on a floating wooden framework (a raft); sometimes netting is used instead of ropes; in some cases the seaweed is simply placed on the bottom of a pond and not fixed in any way; in more open waters, one kind of seaweed is either forced into the soft sediment on the sea bottom with a fork-like tool, or held in place on a sandy bottom by attaching it to sand-filled plastic tubes.
>
> Cultivation involving a reproductive cycle, with alternation of generations, is necessary for many seaweeds; for these, new plants cannot be grown by taking cuttings from mature ones. This is typical for many of the brown seaweeds, and *Laminaria* species are a good example; their life cycle involves alternation between a large sporophyte and a microscopic gametophyte-two generations with quite different forms. The sporophyte is what is harvested as seaweed, and to grow a new sporophyte it is necessary to go through a sexual phase involving the gametophytes. The mature sporophyte releases spores that germinate and grow into microscopic gametophytes. The gametophytes become fertile, release sperm and eggs that join to form embryonic sporophytes. These slowly develop into the large sporophytes that we harvest. The principal difficulties in this kind of cultivation lie in the management of the transitions from spore to gametophyte to embryonic sporophyte; these transitions are usually carried out in land-based facilities with careful control of water temperature, nutrients and light. The high costs involved in this can be absorbed if the seaweed is sold as food, but the cost is normally too high for production of raw material for alginate production.
>
> Where cultivation is used to produce seaweeds for the hydrocolloid industry (agar and carrageenan), the vegetative method is mostly used, while the principal seaweeds used as food must be taken through the alternation of generations for their cultivation.

1.4 CHEMICAL COMPOSITION

A summary of the chemical composition of various filamentous algae and seaweeds is presented in Table 1.1. Algae are receiving increasing attention as possible alternative protein sources for farmed fish, particularly in tropical developing countries, because of their high protein content (especially the filamentous blue-green algae).

The dry matter basis (DM) analyses reviewed in Table 1.1 show that the protein levels of filamentous blue green algae ranged from 60–74 percent. Those for filamentous green algae were much lower (16–32 percent). The protein contents of green and red seaweeds were quite variable, ranging from 6–26 percent and 3–29 percent respectively. The levels reported for *Eucheuma/ Kappaphycus* were very low, ranging from 3–10 percent, but the results for *Gracilaria*, with one exception, were much higher (16–20 percent). The one analysis for *Porphyra* indicated that it had a protein level (29 percent) comparable to filamentous green algae. Some information on the amino acid content of various aquatic macrophytes is contained in Annex 1.

The lipid levels reported for *Spirulina* (Table 1.1), with one exception (Olvera-Novoa *et al.* (1998), were between and 4 and 7 percent. Those for filamentous green algae varied more widely (2–7 percent). The lipid contents of both green (0.3–3.2 percent) and red seaweeds (0.1–1.8 percent) were generally much lower than those of filamentous algae. The ash content of filamentous blue-green algae ranged from 3–11 percent but those of filamentous green algae were generally much higher, ranging from just under 12 percent to one sample of *Cladophora* that had over 44 percent. The ash contents of green seaweeds ranged from 12–31 percent. Red seaweeds had an extremely wide range of ash contents (4 to nearly 47 percent) and generally had higher levels than the other algae shown in Table 1.1.

1.5 USE AS AQUAFEED

Several feeding trials have been carried out to evaluate algae as fish feed. Algae have been used fresh as a whole diet and dried algal meal has been used as a partial or complete replacement of fishmeal protein in pelleted diets.

1.5.1 Algae as major dietary ingredients

A summary of the results of selected growth studies on the use of fresh algae or dry algae meals as major dietary ingredients for various fish species and one marine shrimp is presented in Table 1.2. Dietary inclusion levels in these studies varied from 5 to 100 percent. Fishmeal-based dry pellets or moist diets were used as control diets.

The results of the earlier growth studies showed that the performances of fish fed diets containing 10–20 percent algae or seaweed meal were similar to those fed fishmeal based standard control diet. The responses were apparently similar for most of the fish species tested. These inclusion levels effectively supplied only about 3–5 percent protein of the control diet. In most cases, these control diets contained about 26–47 percent crude protein. This shows that only about 10–15 percent of dietary protein requirement can be met by algae without compromising growth and food utilization. There was a progressive decrease in fish performance when dietary incorporation of algal meal rose above 15–20 percent. However, although reduced growth responses were recorded with increasing inclusion of algae in the diet, the results of feeding trials with filamentous green algae for *O. niloticus* and *T. zillii* indicated that SGR of 60–80 percent of the control diet could be achieved with dietary inclusion levels as high as 50–70 percent.

Recent work by Kalla *et al.* (2008) appears to indicate that the addition of *Porphyra* spheroplasts to a semi-purified red seabream diet improved SGR. In addition, Valente *et al.* (2006) recorded improvements in SGR when dried *Gracilaria busra-pastonis* replaced 5 or 10 percent of a fish protein hydrolysate diet for European seabass.

TABLE 1.1
Chemical analyses of some common algae and seaweeds

Algae/ seaweed	Moisture (percent)	Proximate composition[1] (percent DM)					Minerals[1] (percent DM)		Reference
		CP	EE	Ash	CF	NFE	Ca	P	
Filamentous blue-green algae									
Spirulina maxima, spray dried powder	6.0	63.8	5.3	9.6	n.s.		n.s.	n.s.	Henson (1990)
Spirulina, commercial dry powder	3-6	60.0	5.0	7.0	7.0	21.0	n.s.	n.s.	Habib et al. (2008)
Spirulina spp., dry powder	n.s.	55-70	4-7	3-11	3-7		n.s.	n.s.	Habib et al. (2008)
Spirulina maxima, dry powder, Mexico	10.2	73.7	0.7	10.5	2.1	13.0	n.s.	n.s.	Olvera-Nova et al. (1998)
Filamentous green algae									
Spirogyra spp., fresh, USA	95.2	17.1	1.8	11.7	10.0[2]		n.s.	n.s.	Boyd (1968)
Cladophora glomerata, meal, Scotland	1.6	31.6	5.2	23.6	11.2	28.4	n.s.	n.s.	Appler and Jauncey (1983)
Cladophora sp., fresh, USA[3]	90.5	15.8	2.1	44.3	13.3	24.5[4]	n.s.	n.s.	Pine, Anderson and Hung (1989)
Hydrodictyon reticulatum, fresh, USA	96.1	22.8	7.1	11.9	18.1[2]		n.s.	n.s.	Boyd (1968)
Hydrodictyon reticulatum, meal, Belgium	5.7	27.7	1.9	32.6	14.9	22.9	n.s.	n.s.	Appler (1985)
Green seaweeds									
Ulva reticulata, fresh, Tanzania	n.s.	25.7	n.s.	18.3	38.5		n.s.	0.1	Msuya and Neori (2002)
Ulvaria oxysperma, dried, Brazil	16-20	6-10	0.5-3.2	17-31	3-12		n.s.	n.s.	Pádua, Fontoura and Mathias (2004)
Ulva lactuca, dried, Brazil	15-18	15-18	1.2-1.8	12-13	9-12		n.s.	n.s.	Pádua, Fontoura and Mathias (2004)
Ulva fascita, dried, Brazil	18-20	13-16	0.3-1.9	17-20	9-11		n.s.	n.s.	Pádua, Fontoura and Mathias (2004)
Red seaweeds									
Eucheuma cottonii, fresh, Indonesia	91.3	4.9	0.4	43.5	8.4	42.8[4]	0.5	0.2	Tacon et al. (1990)
Eucheuma cottonii, dry powder, Malaysia	10.6	9.8	1.1	46.2	5.9	37.0[4]	0.3	n.s.	Matanjun et al. (2009)
Eucheuma denticulatum, fresh, Tanzania	n.s.	7.6	n.s.	46.6	22.3		n.s.	<0.1	Msuya and Neori (2002)
Kappaphycus alvarezii, oven dried meal, Philippines	10.1	3.2	0.6	18.1	5.9	72.2[4]	n.s.	n.s.	Peñaflorida and Golez (1996)
Gracilaria heteroclada, oven dried meal, Philippines	9.3	17.3	1.8	21.7	4.6	54.6	n.s.	n.s.	Peñaflorida and Golez (1996)
Gracilaria lichenoides, fresh, Indonesia	88.1	15.6	1.2	36.7	6.6	39.9[4]	0.8	0.3	Tacon et al. (1990)
Gracilaria sp., sun-dried meal, Thailand	7.2	19.9	0.1	31.4	4.9	43.7	n.s.	n.s.	Briggs and Funge-Smith (1996)
Gracilaria crassa, fresh, Tanzania	n.s.	13.2	n.s.	15.0	38.7		n.s.	<0.1	Msuya and Neori (2002)
Porphyra purpurea, meal, England	4.7	28.7	0.4	4.1	6.7	60.1[4]	n.s.	n.s.	Davies, Brown and Camilleri (1997)

DM = dry matter; CP = crude protein; EE = ether extract; CF = crude fibre; NFE = nitrogen free extract; Ca = calcium; P = phosphorus
[1] Mean of proximate composition values of algae collected from flowing and static water
[2] Cellulose
[3] Adjusted or calculated; not as cited in original publication
[4] Adjusted or calculated; not as cited in original publication

TABLE 1.2
Performance of various fish species fed fresh algae or dried algal meal

Algae/ fish species	Rearing system	Rearing days	Control diet	Composition of test diet	Inclusion level (percent)	Fish size (g)	SGR (percent)	SGR as percent of control	FCR	References
Filamentous green algae										
Cladophora glomerata/ Nile tilapia (Oreochromis niloticus)	Laboratory recirculatory system	56	Fish meal based pellet (30 percent protein)	5, 10, 15 and 20 percent protein of control feed replaced by algal meal and one diet prepared by algal meal as the only sources of protein (25 percent protein)	16.1	1.88-2.09	3.11	97.5	1.21	Appler and Jauncey (1983)
					32.3		2.80	87.8	1.42	
					48.4		2.77	86.8	1.51	
					64.5		2.06	64.6	2.09	
					82.5		1.85	58.0	2.33	
Hydrodictyon reticulatum/ Nile tilapia (Oreochromis niloticus)	Laboratory recirculatory system	50	Fish meal based pellet (30 percent protein)	5, 10, 15 and 20 percent protein of control feed replaced by algal meal and one diet prepared by algal meal as the only sources of protein (25 percent protein)	19.2	0.92-1.04	2.22	91.7	1.83	Appler (1985)
					38.3		1.85	76.4	2.18	
					57.5		1.48	61.2	2.49	
					70.6		1.52	62.8	2.63	
					98.5		1.07	44.2	3.60	
Hydrodictyon reticulatum/ Redbelly tilapia (Tilapia zillii)	Laboratory recirculatory system	50	Fishmeal based pellet (30 percent protein)	5, 10, 15 and 20 percent protein of control feed replaced by algal meal and one diet prepared by algal meal as the only sources of protein (25 percent protein)	19.2	0.91-1.16	2.04	107.9	2.09	Appler (1985)
					38.3		1.73	91.5		
					57.5		1.45	76.7		
					70.6		1.44	76.2		
					98.5		1.05	55.6		
Filamentous blue-green algae										
Spirulina/ Java tilapia (Oreochromis mossambicus)	Indoor static tank	25	Fishmeal based moist diet (26 percent protein)	11 percent fishmeal replaced by Spirulina meal	11.0	7.4-8.3	1.96	101.0	-	Chow and Woo (1996)
Seaweeds										
Porphyra purpurea/ thick-lipped grey mullet (Chelon labrosus)	Flow-through system	70	Fishmeal based pellet (47 percent protein)	4.5 and 9.0 percent protein of control feed replaced by seaweed meal	16.5	1.15	2.65	88.6	2.06	Davies, Brown and Camilleri (1997)
					33.0	1.15	2.47	82.6	2.28	

TABLE 1.2 (cont.)
Performance of various fish species fed fresh algae or dried algal meal

Algae/ fish species	Rearing system	Rearing days	Control diet	Composition of test diet	Inclusion level (percent)	Fish size (g)	SGR (percent)	SGR as percent of control	FCR	References
Porphyra sp./ Red seabream (Pagrus major)	Flow-through system	42	Fishmeal based semi-purified diet (51 percent protein)	5 percent Porphyra spheroplasts added to diet	5.0	15.4	3.47	111.6	1.52	Kalla et al. (2008)
Ulva rigida/ European seabass (Dicentrarchus labrax)	Recirculation system	70	Fish protein hydrolysate based diet (60.8 percent protein)	5 and 10 percent fish protein hydrolysate replaced by dried seaweed	5.0 10.0	4.7 4.7	2.63 2.54	89.8 86.7	1.68 1.80	Valente et al. (2006)
Gracilaria cornea/ European seabass (Dicentrarchus labrax)	Recirculation system	70	Fish protein hydrolysate based diet (60.8 percent protein)	5 and 10 percent fish protein hydrolysate replaced by dried seaweed	5.0 10.0	4.7 4.7	2.63 1.78	89.8 60.8	1.74 2.31	Valente et al. (2006)
Gracilaria busra-pastonis/ European seabass (Dicentrarchus labrax)	Recirculating system	70	Fish protein hydrolysate based diet (60.8 percent protein)	5 and 10 percent fish protein hydrolysate replaced by dried seaweed	5.0 10.0	4.7 4.7	2.98 3.37	101.7 115.0	1.56 1.48	Valente et al. (2006)
Gracilaria lichenoides/ rabbitfish (Siganus canaliculatus)	Floating net cages	100	Carp starter pellet (27 percent protein)	Fresh live seaweed was fed as sole diet	100.0	50.1	Negative growth displayed. SGR of control 0.63 percent			Tacon et al. (1990)
Eucheuma cottonii/ rabbitfish (Siganus canaliculatus)	Floating net cages	100	Carp starter pellet (27 percent protein)	Fresh live seaweed was fed as sole diet	100.0	48.8	Negative growth displayed. SGR of control 0.63 percent			Tacon et al. (1990)
Gracilaria sp./ Giant tiger prawns (Penaeus monodon)	Brackishwater recirculatory system	60	Soybean and fish meal based diet (35 percent protein)	1, 2, 3 and 6 percent protein of control feed replaced by seaweed meal. Seaweed meal incorporated by replacing soybean meal and wheat flour	5.0 10.0 15.0 30.0	0.024	7.88 8.03 7.88 7.33	98.3 100.1 98.3 91.4	3.33 3.35 3.50 4.14	Briggs and Funge-Smith (1996)

TABLE 1.3
Results of investigations on the use of algae as additives in fish feed

Algae[1]	Inclusion level (percent)	Fish species	Effect	References
Blue-green algae				
Spirulina	2.0	Red sea bream	Improved carcass quality through modification of muscle lipids	Mustafa et al. (1994a)
	2.0	Red sea bream	Improved muscle quality; increased firmness and robustness of raw meat; and improved growth and protein synthetic activity	Mustafa, Umino and Nakagawa (1994)
	5.0	Red sea bream	Elevated growth rates; improved feed conversion, protein efficiency and muscle protein deposition	Mustafa et al. (1994b)
	5.0	Nibbler	Improved growth	Nakazoe et al. (1986)
	...	Striped jack	Improved flesh texture and taste	Liao et al. (1990)
	2.5	Cherry salmon	Elevated growth rates, bright skin colour and fin appearance; improved flavour and firm flesh	Hensen (1990)
	0.5	Yellowtail	Increased survivability and improved weight gain	Hensen (1990)
Spirulina maxima	20.9 [40.0 replacement of fish meal]	Mozambique tilapia	Final body weight, daily weight gain, SGR, feed intake, PER and apparent nitrogen utilization showed no significant differences with control diet	Olvera-Novoa et al. (1998)
Brown algae				
Ascophyllum nodosum	5.0 & 10.0	Red sea bream	Improved growth and feed efficiency at 5 percent inclusion level	Yone, Furuichi and Urano (1986a)
	5.0	Red sea bream	Delayed absorption of dietary carbohydrate and protein. The dietary nutrients are utilized effectively by this delaying effect of the seaweed; thus the growth and feed efficiency of red sea bream are improved	Yone, Furuichi and Urano (1986b)
	5.0	Red sea bream	Elevated growth rates; improved feed conversion, protein efficiency and muscle protein deposition	Mustafa et al. (1994b)
	5.0	Red sea bream	Increased growth, feed efficiency and protein deposition. Elevated liver glycogen and triglyceride accumulation in muscle	Mustafa et al. (1995)
	0.5	Yellowtail	Prevented a nutritional disease that causes retardation of growth and high mortality	Nakagawa et al. (1986)
Undaria pinnatifida	5.0	Rockfish	Showed prominent physiological effects on haematocrit value and red blood cell number	Yi and Chang (1994)
	5.0 & 10.0	Red sea bream	Improved growth and feed efficiency, and higher muscle lipid deposition at 5 percent level of inclusion	Yone, Furuichi and Urano (1986a)

TABLE 1.3 (cont.)
Results of investigations on the use of algae as additives in fish feed

Algae[1]	Inclusion level (percent)	Fish species	Effect	References
Red algae				
Porphyra yezoensis	5.0	Red sea bream	Increased growth, feed efficiency and protein deposition. Elevated liver glycogen and triglyceride accumulation in muscle	Mustafa et al. (1995)
Porphyra yezoensis	2.0	Yellowtail	Improved flesh quality	Morioka et al. (2008)
Porphyra spheroplasts	5.0	Red sea bream	Survival, growth and nutrient retention significantly higher than control	Kalla et al. (2008)
Green algae				
Ulva conglobata	5.0	Nibbler	Improved growth	Nakazoe et al. (1986)
Ulva pertusa	2.5, 5.0, 10.0 & 15.0	Black sea bream	Ulva meal diets repressed lipid accumulation in intraperitoneal body fat without loss of growth and feed efficiency. Fish fed 2.5, 5 and 10 percent Ulva meal did not show significant body weight loss during wintering. During starvation, lipid reserves were preferentially mobilized for energy	Nakagawa et al. (1993)
Ulva pertusa extract	10.0	Black sea bream	Improved tolerance to hypoxia	Nakagawa et al. (1984)
Ulva pertusa	5.0	Red sea bream	Activated lipid mobilization and suppressed protein breakdown observed during starvation for fish fed *Ulva* meal supplemented diet before starvation. Preferential use of glycogen observed	Nakagawa and Kasahara (1986)
	5.0	Red sea bream	Demonstrated a decrease in susceptibility to *Pasteurella piscicida*, an elevation of phagocytosis and spontaneous haemolytic and bactericidal activity	Satoh, Nakagawa and Kasahara (1987)
	5.0	Red sea bream	Increased growth, feed efficiency and protein deposition. Elevated liver glycogen and triglyceride accumulation in muscle	Mustafa et al. (1995)

[1] Algae were added as dried meal in all diets except otherwise stated

However, the conclusions of the latter authors are confused by the fact that the test diets were not iso-nitrogenous with the control diet; in fact test diets had a lower protein level.

Total replacement of fishmeal by algal meal showed very poor growth responses for *O. niloticus* (Appler and Jauncey, 1983; Appler, 1985) and *T. zillii* (Appler, 1985). Appler and Jauncey (1983) recorded a SGR of 58 percent of control diet when the filamentous green alga (*Cladophora glomerata*) meal was used as the sole source of protein for Nile tilapia. Similarly, Appler (1985) recorded SGRs of 44 percent and 56 percent of control diets when the filamentous green alga (*Hydrodictyon reticulatum*) meal was used as the sole source of protein for *O. niloticus* and *T. zillii*.

Tacon *et al.* (1990) used fresh live seaweeds (*Gracilaria lichenoides* and *Eucheuma cottonii*) as the total diet for rabbitfish in net cages. In both cases negative growth was displayed, although the daily feed intake was more than the control diet. On a dry matter basis, the daily feed intake was 1.99 and 1.98 g/fish/day respectively for *E. cottonii* and *G. lichenoides*, while the feed intake for carp pellets (control diet) was 1.80 g/fish/day. Apparently, a good feeding response was observed for both the seaweeds but very poor feed efficiency was displayed. Apart from commonly observed impaired growth, the use of algae as the sole source of protein in fish feed can also result in malformation (Meske and Pfeffer, 1978).

The apparently poor performance of fish fed diets containing higher inclusion levels of algae may be attributable to several factors. Appler (1985) observed that most of the aquatic plants including algae contain 40 percent or more of carbohydrate, of which only a small fraction consists of mono- and di-saccharides. Low digestibility of plant materials has been attributed to a preponderance of complex and structural carbohydrates. The poor digestibility and the subsequent poor levels of utilization obtained for both tilapia species with increased dietary algal levels may thus be attributable in part to the presence of indigestible algal materials. Pantastico, Baldia and Reyes (1985) reported that newly hatched Nile tilapia fry (mean weight 0.7 mg) did not survive at all when unialgal cultures of *Euglena elongata* and *Chlorella ellipsoidea* were fed to them. These authors concluded that the mortality of tilapia fry might be due to factors such as toxicity and cell-wall composition of the algae fed. This phenomenon might also be attributed to poor digestion of plant material by the less developed digestive system of newly hatched larva. In contrast, Chow and Woo (1990) recorded significantly higher gut cellulase activity in *O. mossambicus* fed *Spirulina*, indicating the ability of this tilapia species to digest cellulose, the main constituent of plant cell walls. Ayyappan *et al.* (1991) conducted a *Spirulina* feeding experiment with carp species. The fry stage of catla (*Catla catla*), rohu (*Labeo rohita*), mrigal (*Cirrhinus mrigala*), silver carp (*Hypophthalmicthys molitrix*), grass carp (*Ctenopharyngodon idella*) and common carp (*Cyprinus carpio*) were fed with an experimental diet in which 10 percent dried *Spirulina* powder was added to a 45:45 mixture of rice bran and groundnut oil cake. A 50:50 bran-groundnut oil cake control diet was used. The mean specific growth rates of fish fed on the two diets were: catla 0.17, 0.27; rohu 0.19, 0.63; mrigal 0.54, 0.73; grass carp 0.02, 0.40; and common carp 0.15, 0.20; with significant differences between the treatments ($F_{1,4} = 8.88$; $P < 0.05$) and fish species ($F_{4,4} = 5.03$; $P < 0.10$). Rohu and mrigal showed significantly ($P < 0.05$) higher SGRs than catla and common carp. These results clearly demonstrated the beneficial effect of the *Spirulina* diet on the yield and quality of carp fry.

Dietary supplementation of *Chlorella ellipsoidea* powder at 2 percent on a dry-weight basis showed higher weight gain and improved feed efficiency and protein efficiency ratios in juvenile Japanese flounders (*Paralichthys olivaceus*); the addition of *Chlorella* had positive effects as it significantly reduced serum cholesterol and body fat levels and also led to improved lipid metabolism (Kim *et al.*, 2002).

Clearly, no definite conclusions can be arrived at this stage about the value of using macroalgae as major dietary ingredients or protein sources in aquafeeds. Moderate growth responses and good food utilization (FCR 1.5–2.0) were generally recorded when dried algal meal were used as a partial replacement of fishmeal protein. However, the collection, drying and pelletization of algae require considerable time and effort. Furthermore, cultivation costs would have to be taken into consideration. Therefore, further cost-benefit on-farm trials that take these costs into consideration are needed before any definite conclusions on the future application of algae as fish feed can be drawn.

1.5.2 Algae as feed additives

The main applications of microalgae for aquaculture are associated with nutrition, being used fresh (as sole component or as food additive to basic nutrients) for colouring the flesh of salmonids and for inducing other biological activities (Muller-Feuga, 2004). Several investigations have been carried out on the use of algae as additives in fish feed. Feeding trials were carried out with many fish species, most commonly red sea bream (*Pagrus major*), ayu (*Plecoglossus altivelis*), nibbler (*Girella punctata*), striped jack (*Pseudoceranx dentex*), cherry salmon (*Oncorhynchus masou*), yellowtail (*Seriola quinqueradiata*), black sea bream (*Acanthopagrus schlegeli*), rainbow trout (*Oncorhynchus mykiss*), rockfish (*Sebastes schlegeli*) and Japanese flounder (*Paralichthys olivaceus*). Various types of algae were used; the most extensively studied ones have been the blue-green algae *Spirulina* and *Chlorella*; the brown algae *Ascophyllum*, *Laminaria* and *Undaria*; the red alga *Porphyra*; and the green alga *Ulva*. Fagbenro (1990) predicted that the incidence of cellulase activity could be responsible for the capacity of the catfish *Clarias isherencies* to digest large quantities of Cyanophyceae.

A summary of the results of selected feeding trials with algae as feed additives is presented in Table 1.3. Most of these research studies were conducted in Japan with Japanese fish species, although the results may well be applicable to other species and in other countries.

Table 1.3 shows that dried algal meals or their extracts have been added to test fish diets at levels up to 21 percent level. The responses of test fish fed algae supplemented diets were compared with fish fed standard control diets. Although various types of algae and fish species were used in these evaluations, not all algae were evaluated as feed additives for every different species. As the main biochemical constituents and digestibility are different among algae, the effect of dietary algae varies with the algae and fish species (Mustafa and Nakagawa, 1995). While studying the effect of two seaweeds (*Undaria pinnatifida* and *Ascophyllum nodosum*) at different supplementation levels for red sea bream, Yone, Furuichi and Urano (1986a) observed best growth and feed efficiency from a diet containing 5 percent *U. pinnatifida* followed by a diet containing 5 percent *A. nodosum*. Similarly, Mustafa *et al.* (1994b) observed more pronounced effects on growth and feed utilization of red sea bream by feeding a diet containing *Spirulina* compared to one containing *Ascophyllum*. In another study, Mustafa *et al.* (1995) studied the comparative efficacy of three different algae (*Ascophyllum nodosum*, *Porphyra yezoensis* and *Ulva pertusa*) for red sea bream and noted that feeding *Porphyra* showed the most pronounced effects on growth and energy accumulation, followed by *Ascophyllum* and *Ulva*. However, research results obtained so far do not specifically identify any specific algae as the most suitable as feed additives for any particular fish species.

Nevertheless, the results of various research studies show that algae as dietary additives contribute to an increase in growth and feed utilization of cultured fish due to efficacious assimilation of dietary protein, improvement in physiological activity, stress response, starvation tolerance, disease resistance and carcass quality. In fish fed algae-supplemented diets, accumulation of lipid reserves was generally well controlled and the reserved lipids were mobilized to energy prior to muscle protein degradation in response to energy requirements. In complete pelleted diets, algal supplementation of 5 percent or less was found to be adequate.

Spirulina are widely used as feed additives in the Japanese fish farming industry. Henson (1990) reported that *Spirulina* improved the performances of ayu, cherry salmon, sea bream, mackerel, yellowtail and koi carp. The levels of supplementation used by Japanese farmers are 0.5-2.5 percent. Henson (1990) further reported that Japanese fish farmers used about US$2.5 million worth of *Spirulina* in 1989. Five important benefits reported by using a feed containing this alga were improved growth rates; improved carcass quality and colouration; higher survival rates; reduced requirement for medication; and reduced wastes in effluents. However, the high cost of most of these algae may limit their use to the commercial production of high value fish only.

2. Floating aquatic macrophytes – *Azolla*

Floating aquatic macrophytes are defined as plants that float on the water surface, usually with submerged roots. Floating species are generally not dependent on soil or water depth.

Azolla spp. are heterosporous free-floating freshwater ferns that live symbiotically with *Anabaena azollae*, a nitrogen-fixing blue-green algae. These plants have been of particular interest to botanists and Asian agronomists because of their association with blue-green algae and their rapid growth in nitrogen deficient habitats (Islam and Haque, 1986). The genus *Azolla* includes six species distributed widely throughout temperate, sub-tropical and tropical regions of the world. It is not clear whether the symbiont is the same in the various *Azolla* species.

Azolla spp. consists of a main stem growing at the surface of the water, with alternate leaves and adventitious roots at regular intervals along the stem. Secondary stems develop at the axil of certain leaves. *Azolla* fronds are triangular or polygonal and float on the water surface individually or in mats. At first glance, their gross appearance is little like what are conventionally thought of as ferns; indeed, one common name for them is duckweed ferns. Plant diameter ranges from 1/3 to 1 inch (1-2.5 cm) for small species like *Azolla pinnata* to 6 inches (15 cm) or more for *A. nilotica* (Ferentinos, Smith and Valenzuela, 2002).

2.1 CLASSIFICATION

The genus *Azolla* belongs to the single genus family Azollaceae. The six recognizable species within the genus are grouped under two subgenera: *Euazolla* and *Rhizosperma*.

The four species under the sub-genus *Euazolla* are *A. filiculoides*, *A. caroliniana*, *A. mexicana* and *A. microphylla*. It is thought that these four species originated from temperate, sub-tropical and tropical regions of North and South America (Van Hove, 1989). However, Zimmerman *et al.* (1991) found three of these species (*A. caroliniana*, *A. mexicana* and *A. microphylla*) to have close taxonomic affinity and similar responses to phosphorus deficiency and recommended that they be considered as a single species.

The two species under sub-genus *Rhizosperma* are *A. nilotica* and *A. pinnata*. *A. nilotica* is a native of East Africa and can be found from Sudan to Mozambique (Van Hove, 1989). *A. pinnata* has two different varieties that vary in their distribution patterns. *A. pinnata* var. *imbricata* originates from subtropical and tropical Asia while *A. pinnata* var. *pinnata* occurs in Africa and is known as African strain.

2.2 CHARACTERISTICS
2.2.1 Importance

Because *Azolla* has a higher crude protein content (ranging from 19 to 30 percent) than most green forage crops and aquatic macrophytes and a rather favourable essential amino acid (EAA) composition for animal nutrition (rich in lysine), it has also attracted the attention of livestock, poultry and fish farmers (Cagauan and Pullin, 1991). In Asia *Azolla* has been long used as green manure for crop production and a supplement to diets for pigs and poultry. Some strains of *Azolla* can fix as much as 2-3 kg of nitrogen/ha/day. *Azolla* doubles its biomass in 3-10 days, depending on conditions, and easily reaches a standing crop of 8-10 tonnes/ha fresh weight in Asian rice fields; 37.8 tonnes/ha fresh weight (2.78 tonnes/ha dry weight) has been reported for *A. pinnata* in India (Pullin and Almazan, 1983). Recently, Liu *et al.* (2008) have reported the application of *Azolla* as a controlled ecological life support system (CELSS) for its strong photosynthetic oxygen-releasing capacity. *Azolla* provides a protected environment and a fixed source of carbon to the blue-green filamentous algae *Anabaena* spp. (Wagner, 1997).

2.2.2 Environmental requirements

The potential for rearing *Azolla* is restricted by climatic factors, water and inoculum availability, the incidence of pests, phosphorus requirements and the need for labour intensive management (Cagauan and Pullin, 1991). Water is the fundamental requirement for the growth and multiplication of *Azolla*. The plant is extremely sensitive to lack of water. Although *Azolla* can grow on wet mud surfaces or wet pit litters, it prefers growing in a free-floating state (Becking, 1979). A strip of water not more than a few centimetres deep favours growth because it provides good mineral nutrition, with the roots not too far from the soil, and also because it reduces wind effects (Van Hove, 1989). Strong winds can accumulate *Azolla* to one side of the stretch of water, creating an overcrowded condition and thus slowing growth.

Azolla can survive a water pH ranging from 3.5-10, reported optimum growth occurring at pH 4.5-7.0. Watanabe *et al.* (1977) reported that the growth of *Azolla* was optimum at pH 5.5 and FAO (1977) recorded that soils of pH 6 to 7 support the best growth.

Salinity tolerance of *Azolla* species also varies. The growth rate of *A. pinnata* declines as its salinity increases above 380 mg/l (Thuyet and Tuan, 1973). According to Reddy *et al.* (2005) *Azolla* can withstand salinity of up to 10 ppt but Haller, Sutton and Burlowe (1974) reported that the growth of *A. caroliniana* ceases at about 1.3 ppt and higher concentrations result in death. *A. filiculoides* has been reported to be most salt-tolerant (I. Watanabe pers. comm., cited by Cagauan and Pullin, 1991).

Azolla grows in full to partial shade (100-50 percent sunlight) with growth decreasing quickly under heavy shade (Ferentinos *et al.*, 2002). Generally, *Azolla* requires 25-50 percent full sunlight for its normal growth; slight shade is of benefit to its growth in field conditions. The biomass production greatly decreases at a light intensity lower than 1 500 lux (Liu *et al.*, 2008).

Like all other plants, *Azolla* needs all the macro- and micro-nutrients for its normal growth and vegetative multiplication. All elements are essential; phosphorus is often the most limiting element for its growth. For normal growth, 0.06 mg/l/day is required. This level can be achieved in field conditions by the split application of superphosphate at 10-15 kg/ha (Sherief and James, 1994). 20 mg/l is the optimum concentration (Ferentinos *et al.*, 2002). The symptoms of phosphorous deficiency are red-coloured fronds (due the presence of the pigment anthocyanin), decreased growth and curled roots. Macronutrients such as P, K, Ca and Mg and micronutrients such as Fe, Mo and Co have been shown to be essential for the growth and nitrogen fixation of *Azolla* (Khan and Haque, 1991).

The temperature tolerance of *Azolla* varies widely between its species and strains. In general, *Azolla* has low tolerance to high temperature and that restricts its use in tropical agriculture. There are, however, strains that can adapt successfully to high temperature. Cagauan and Pullin (1991) ranked different *Azolla* species from the most to the least heat-tolerant, based on the optimum temperature for good growth: *A. mexicana* > *A. pinnata* var. *pinnata* > *A. microphylla* > *A. pinnata* var. *imbricata*, *A. caroliniana* > *A. filiculoides* (Table 2.1). In general, the optimum temperature for growth of all *Azolla* species is around 25 °C, except that of *A. mexicana*, whose optimum temperature is

TABLE 2.1
Temperature tolerance of five species of *Azolla*

Subgenera	Species	Water temperature (°C)		
		Minimum	Maximum	Optimum for growth
Euazolla	A. filiculoides	0-10	38-42	20-25
	A. caroliniana	<0-10	45	20-30
	A. mexicana	-	-	30-33
	A. microphylla	5-8	45	25-30
Rhizosperma	A. pinnata			
	A. pinnata var. pinnata	<5	>40	16-33
	A. pinnata var. imbricata	0	45	20-30

Source: modified from Cagauan and Pullin (1991)

~30 °C. According to Sherief and James (1994), the favourable water temperature for rapid multiplication of *Azolla* is generally between 18 and 26 °C.

The optimum relative humidity for *Azolla* growth is between 85-90 percent. *Azolla* becomes dry and fragile at a relative humidity lower than 60 percent.

2.3 PRODUCTION

Multiplication of *Azolla* in nature and in the laboratory is entirely through vegetative reproduction. However, sexual reproduction, which is essential to the survival of the population during temporary adverse conditions also, occurs. When *Azolla* fronds reach a certain size depending on the species and the environment, generally 1 to 2 cm in diameter, the older secondary stems detach themselves from the main stem as a result of the formation of an abscission layer, thus giving rise to new fronds. This is the most usual mode of multiplication.

Sherief and James (1994) have described a simple *Azolla* nursery method for its large-scale multiplication in the field for Indian farmers. The field for an *Azolla* nursery must be thoroughly prepared and levelled uniformly. It is divided into different plots by providing suitable bunds and irrigation channels. Water is manipulated at a depth of 10 cm. Ten kg of fresh cattle dung mixed in 20 L of water is sprinkled in each plot and an *Azolla* inoculum of 8 kg is introduced to each plot. Superphosphate (100 g) is applied in three split doses at intervals of four days as a top dressing fertilizer. For insect control, furadone granules at 100 g/plot are applied seven days after inoculation. Fifteen days after inoculation, *Azolla* is harvested. From one harvest, 40-55 kg of fresh *Azolla* is obtained from each plot. Reddy and DeBusk (1985) reported the yield of *Azolla* (*A. caroliniana*) to be 10.6 t DM/ha/year in nutrient non-limiting waters of central Florida, USA.

According to Ferentinos *et al.* (2002) the nitrogen fixation capacity of *Azolla* was found to vary from 53-1 000 kg/ha with a dry matter production of 39-390 tonnes/ha, in crop cycles of 40-365 days. The linear growth phase is usually between 6 and 21 days and is characterized by low lignin and cell wall fractions. Due to its high lignin content (20 percent), nitrogen is released slowly from the plant initially, with about two-thirds released on the first 6 weeks after application. Under flooded conditions, 40-60 percent of the available N is released after 20 days and 55-90 percent within 40 days after application

Reddy *et al.* (2005) described the production of *Azolla* in earthen raceways (10.0 m x 1.5 m x 0.3 m) in CIFA, Bhubaneswar. 6 kg of *Azolla* was inoculated in each raceway. 50 kg single super phosphate and pesticide (1-2 mg/l) were applied and a water depth of 5-10 cm was maintained. 18-24 kg/raceway/week was produced. About one tonne of *Azolla* could be harvested every week from water spread area of 650 m^2, with a phosphorus input-nitrogen output ratio of 1:4.8.

TABLE 2.2
Chemical analyses of various Azolla species

Azolla species	Moisture (percent)	Composition[1] (percent DM[1])					Minerals (percent DM)			Reference
		CP	EE	CF	Ash	CC	Ca	P	K	
A. filiculoides	93.5[2]	25.0-28.5	3.1	n.s.	17.3	4.4-11.5	0.5-1.5	1.0-1.5	6.0	modified from Cagauan and Pullin (1991)
A. caroliniana		20.6-22.6	n.s.	n.s.	n.s.	8.5	0.6	1.3	5.3	modified from Cagauan and Pullin (1991)
A. pinnata var. imbricate		26.0	n.s.	n.s.	n.s.	4.1	0.4	1.3	4.5	modified from Cagauan and Pullin (1991)
A. pinnata (tank culture)		18.2	1.3	n.s.	21.7	n.s.	1.6	0.6	n.s.	modified from Cagauan and Pullin (1991)
A. pinnata (field culture)		22.2	2.9	n.s.	18.3	14.7	n.s.	n.s.	n.s.	modified from Cagauan and Pullin (1991)
A. pinnata		21.4	2.7	12.7	16.2					Alalade and Iyayi (2006)
A. microphylla (lab. culture)		21.8	2.9	n.s.	21.6	15.6	n.s.	n.s.	n.s.	modified from Cagauan and Pullin (1991)
A. microphylla (field culture)		20.0-26.0	3.0-3.5	n.s.	14-15	4.0-14.0	0.7	1.6	5.5	modified from Cagauan and Pullin (1991)
A. microphylla hybrid (field culture)		19.0	4.0-4.5	n.s.	16.0-17.0	2.5-3.0	n.s.	n.s.	n.s.	modified from Cagauan and Pullin (1991)
Various Azolla spp.		13.0-30.0	4.4-6.3	n.s.	9.7-23.8	5.6-15.2	0.2-0.7	0.1-1.6	0.3-6.0	Reddy et al. (2005)
Azolla sp.		n.s.	n.s.	n.s.	n.s.	n.s.	1.0	0.4	2.5	Ferentinos, Smith and Valenzuela, (2002)

[1] CP = crude protein; EE = ether extract; CC = crude cellulose; Ca = calcium; P = phosphorus; K = potassium
[2] Data obtained from Tacon (1987)

2.4 Chemical composition

The chemical composition of *Azolla* species varies with ecotypes and with the ecological conditions and the phase of growth. The dry matter percentage of different *Azolla* species varies widely and there is little agreement between the published data on this subject: values of 5 to 7 percent can, however, be taken as fair estimates (Van Hove, 1989). A summary of the chemical composition of various *Azolla* species is presented in Table 2.2. Generally, the crude protein content is about 19-30 percent DM basis during the optimum conditions for growth (Peters *et al.*, 1979; Becking, 1979). Under natural conditions, values near 20-22 percent are frequent. The protein contents of *Azolla* species are comparable to or higher than that of most other aquatic macrophytes. Like most of the other aquatic macrophytes, *Azolla* have high ash contents, varying between 14-20 percent. No clear interspecific difference in the crude lipid levels of various *Azolla* species occurs; the value is around 3-6 percent on a DM basis.

Amino acid compositions of *Azolla* spp. are presented in Annex 1 Table 2. Generally, these species are low in methionine but high in lysine (except for *A. pinnata*). *A. microphylla* is richest in all EAA except in methionine. The poorest species with respect to most of the EAA is *A. filiculoides* although lysine and methionine contents in this species are moderate. The EAA composition of *Azolla* species is comparable to that of the aquatic plants commonly used as fish feed ingredients. The lysine and methionine contents of most *Azolla* species appear to be higher than some 'conventional' plant protein sources.

2.5 USE AS AQUAFEED

In spite of its attractive nutritional qualities and relative ease of production in ponds and rice-fields, reports on the use of *Azolla* in aquaculture are extremely limited. The value of *Azolla* as a fish feed is still being studied. Available literature on the use of *Azolla* for this purpose has been reviewed as follows under the headings experimental studies and field studies.

2.5.1 Experimental studies

A few studies have been carried out in aquaria to examine the preference for various *Azolla* species by different cichlid species and a carp hybrid. These tests were carried out using fresh *Azolla*; the results are summarized in Table 2.3. These preference tests indicate that *A. caroliniana* (Figure 2.1) is one of the most preferred species for cichlids.

A number of growth studies have been carried out to evaluate *Azolla* as fish feed under laboratory rearing conditions. Most of these studies were conducted on cichlids and little or no attempt was made to use *Azolla* as a feed for grass carp, a predominantly macrophytophagous feeder. In these studies, *Azolla* was fed either in fresh or dried powdered form as a whole feed or by partially replacing fishmeal in pelleted diets.

Almazan *et al.* (1986) conducted a study where *A. pinnata* was fed to Nile tilapia (*Oreochromis niloticus*) fingerlings and adult males. Fingerlings were fed *Azolla* in fresh, powder, and pellet form, replacing the complete control diet mix from 10 percent to 90 percent. The control diet consisted of 40 percent fishmeal, 40 percent rice bran, 10 percent cornstarch, 9 percent corn meal and 1 percent Afsillin (micronutrient premix). Negative

FIGURE 2.1
Azolla/mosquito fern *(Azolla caroliniana)*

Source: www.msrosenthal.com/Ferns/images/Florida_Images/Azolla_caroliniana.jpg

TABLE 2.3
Preference of *Azolla* spp. by different fish species

Fish species	Preferred *Azolla* species[1]	Reference
Cichlasoma fenestratum	(1) *A. microphylla*	Antoine et al. (1986)
	(2) *A. caroliniana*	
Oreochromis niloticus	(1) *A. filiculoides*	Antoine et al. (1986)
	(2) *A. microphylla*	
	(3) *A. caroliniana*	
O. mossambicus	*A. caroliniana*	Lahser (1967)
O. niloticus	*A. microphylla*	Fiogbé, Micha and Van Hove (2004)
Tilapia rendalli	(1) *A. caroliniana*	Micha et al. (1988)
	(2) *A. pinnata var. pinnata*	
	(3) *A. microphylla*	
	(4) *A. filiculoides*	
Hybrid carp (grass carp x bighead carp)	*A. caroliniana*	Cassani (1981)

[1] *Azolla* species are arranged chronologically for each fish species in order of preference i.e. from most preferred to less preferred

or very slow growth was obtained in all *Azolla*-incorporated diets. A lowering of growth performance and FCRs was observed with increasing *Azolla* incorporation. Adult male tilapia were fed dried *Azolla* pellets or fresh *Azolla ad libitum*. Despite feeding to satiation, tilapia suffered weight loss in a 30-day feeding trial. The experiments were carried out in aquaria. Similarly, Antoine et al. (1986) working with *O. niloticus* and *Cichlasoma melanurum* and Micha et al. (1988) with *O. niloticus* and *Tilapia rendalli* reported poor growth and feed utilization when they were fed *A. microphylla*-based diets. Antoine et al. (1986) and Micha et al. (1988) conducted a 70-day growth trial and fed the target species with three different diets: commercial pellets (34 percent protein); fresh *Azolla* plus 28 percent protein commercial pellets (50:50); and fresh *Azolla* (22 percent protein).

In other studies, El-Sayed (1992; 2008) reported extremely poor performance for *O. niloticus* fingerlings and adults fed diet based on *A. pinnata*. This author incorporated dried *Azolla* powder at 25, 50, 75 and 100 percent replacement of fishmeal protein in a fishmeal-based control diet in a 70-day trial. Fresh *Azolla* as a total diet was also used as a control. Growth and feed utilization efficiency of fish fed with the control diet was significantly higher compared to fish fed with *Azolla*-supplemented diets. The performance of fish was inversely related to the increasing dietary incorporation of *Azolla*. In fish fed with the total *Azolla* (dry/fresh) based diet, the reduction was extremely sharp. Fresh *Azolla*-fed adults started losing weight from the 7th week. Fish fed with fresh plant material had significantly higher moisture content than fish fed with formulated diets. Body protein and lipid levels were negatively correlated with the concentrations of *Azolla* in the diets; ash content showed a positive correlation.

In all the experimental studies examined above (Almazan et al., 1986; Antoine et al., 1986; Micha et al., 1988), fish growth was higher in the conventional control diets than in the *Azolla*-based diets. Fish died or negative growth was recorded when fed exclusively with fresh *Azolla*.

In apparent contrast, Santiago et al. (1988) found that *O. niloticus* fry fed rations containing up to 42 percent *of A. pinnata* outperformed fish fed a fishmeal-based control diet. Growth and feed utilization of *O. niloticus* fry improved with increased dietary inclusion of *Azolla* and the survival was unaffected. These results differed from the studies of Almazan et al. (1986), Antoine et al. (1986) and Micha et al. (1988) and it must be pointed out that Santiago et al. (1988) used a 35 percent protein diet with early fry (11-14 mg). In the other studies, the crude protein level was generally lower and the studies were carried out with advanced fry, fingerling or adults. El-Sayed (2008) noted that young Nile tilapia utilized *Azolla* more efficiently than adults.

Fiogbé, Micha and Van Hove (2004) obtained quite favourable results with *Azolla*-based diets fed to juvenile *Oreochromis niloticus* grown in a recirculating system. Six diets were formulated with almost isonitrogenous levels of protein

(27.25-27.52 percent DM) but different levels of dry *Azolla* meal (0, 15, 20, 30, 40 and 45 percent). All diets with incorporated *Azolla* meal showed weight gain. The *Azolla*-free diet and the diet containing 15 percent *Azolla* produced the same growth performance. The least expensive diet, which contained 45 percent *Azolla*, also showed growth and was thought to be potentially useful as a complementary diet for tilapia raised in fertilized ponds. These authors noted that mixing *Azolla* with some agricultural by-products such as rice bran; the use of fermentable by-products such as yeasts; or the addition of purified enzymes; might improve ingestion and digestibility.

Carcass compositions of fish were reported to be markedly affected by feeding with *Azolla*. Antoine *et al.* (1986) observed that when fed with fresh *Azolla*, both *O. niloticus* and *C. melanurum* had higher moisture and lower lipid concentrations. Similar results and an increase in carcass ash content for *O. niloticus* and *T. rendalli* were reported by Micha *et al.* (1988). El-Sayed (1992) also made similar observations when he fed fresh and dried *A. pinnata* to *O. niloticus*. However, his observation differs from the previous authors to the extent that carcass protein content was negatively correlated with *Azolla* levels in the diets, while the other workers recorded no effect on carcass protein content.

The poor growth of fish fed with diets containing higher levels of *Azolla* may be due to excesses or deficiencies of amino acids, according to Fiogbé, Micha and Van Hove (2004). Cole and Van Lunen (1994) found that inadequate levels of essential amino acids resulted in depression of food intake and growth. Deficiencies of one or more amino acids are known to limit protein synthesis, growth or both.

2.5.2 Field studies

Until recently, reports of on-farm utilization of *Azolla* were limited (Cagauan and Pullin, 1991). At that time reports came only from China and Vietnam (Figure 2.2). More recently *Azolla* has increasingly been used as feed and/or fertilizer in studies with rice-fish culture systems in many other Asian countries. Reddy *et al.* (2005) reported that the manuring schedule can be reduced by 30-35 percent through *Anabaena azollae* —*Azolla* biofertilization in aquacultural ponds.

Azolla *in cage culture*

Pantastico, Baldia and Reyes (1986) used fresh whole *A. pinnata* as a supplemental feed for the cage culture of Nile tilapia in Laguna de Bay, Philippines. *Azolla* was propagated in fine mesh net enclosures in the lake and harvested for feeding to tilapia in cages. Four separate experiments were conducted and weight gain was compared with an unfed control. It was assumed that in control cages fish grew by feeding natural food (i.e. plankton) available in the cage. A summary of the results is given in Table 2.4. Although higher weight gain of fish was observed over the unfed control, the difference in mean weight between fish fed fresh *Azolla* and unfed control was about 5-10 g. The results of this cage culture study do not justify fish culture in cages using *Azolla* as the only feed.

FIGURE 2.2
Harvest of fish from a pond (Hoa Binh Province, Viet Nam)

These low-input aquaculture ponds are generally stocked with macrophytophagous fish (primarliy carp species) and fresh *Azolla* (*Azolla pinnata*) are commonly used as supplemental feed.

Courtesy of M.G. Kibria

TABLE 2.4
Cage culture of Nile tilapia using *Azolla* as feed

Initial weight (g)	Stocking density (Numbers/m³)	Duration (months)	Feeding rate (percent)	Fresh *Azolla* Harvest weight (g)	Unfed control Harvest weight (g)
1.3	25	6	35 and 70	30.3 and 36.3	24.7
1.6	50	4	30	75.0	64.1
6.5	100	5	20	20.2	10.9
13.5	150	3	15	29.3	20.2

Rice-fish-Azolla integration

One of the most successful uses of *Azolla* is its use as fertilizer and/or feed in an integrated rice-fish-*Azolla* system. This system is based on convenient layout of the fields, which allows the simultaneous development of rice, *Azolla* and different fish with complementary nutritional requirements (Van Hove, 1989). In this ecological agricultural layout, each of the partners contributes to the equilibrium of the system. The fish (a correct mixture of planktophages, macrophytophages and polypages) derive a benefit from *Azolla* - more or less, depending on the species; their waste promotes the proliferation of plankton that is consumed by some of the fish and fertilizes the rice. The polyphagous fish protect the rice and *Azolla* from a number of insects and molluscan pests.

Cagauan and Pullin (1991) reviewed the rice-fish-*Azolla* integrated system and described its physical set-up, which is provided with pits (pond refuse/ main channel) and ditches (trenches). Lateral or peripheral trenches are interconnected with each other and with the main channel. Trenches serve as links for the fish from the main channel to rice field area and also as a growing area for *Azolla* during the paddy cultivation period. Depending on the size of the rice field, trenches may be dug at 15-20 m intervals in single or rib-shaped patterns. In India, a 0.2 ha rice field was provided with 0.5 m deep and 0.5 m wide trenches and a 1.0 m deep and 1.5 m wide main channel (Shanmugasundaram and Ravi, 1992). Cagauan (1994) used 1 m wide and 0.75 m deep pond refuge connected to an outer trench that was 0.3-0.4 m wide and 0.2-0.3 m deep in a 200 m² paddy field. The trenches and main channels should occupy about 10-15 percent of the rice field area (Cagauan and Pullin, 1991; Shanmugasundaram and Balusamy, 1993). Inoculation of the rice field with *Azolla* at the rate of 4.5-6.0 tonnes/ha is done 20 days before rice transplanting. Propagated *Azolla* biomass is ploughed under, together with inorganic fertilizer, before rice transplanting. The field is then reflooded to allow the *Azolla* that floated during the incorporation to grow and serve as a fish fodder. In case of insufficiency of *Azolla* in the channels and trenches, additional supplemental feed is given. The fish species cultured in these rice-fish-*Azolla* systems are mainly Nile tilapia. Other species are common carp, Indian major carp, Java barb, etc. Grass carp may not be a suitable species for this system, as they may damage the rice crop by feeding on its leaves.

The use of *Azolla* (*A. microphylla*) as a fertilizer in rice-fish farming was studied by Cagauan and Nerona (1986) and Cagauan (1994). Cagauan and Nerona (1986) used three fertilizer regimes: *Azolla* only; inorganic fertilizers (urea and ammonium phosphate) only; and *Azolla* plus inorganic fertilizers for rice-fish culture with Nile tilapia as the target species. When a combination of *Azolla* and inorganic fertilizers was used, it was possible to reduce the standard rate of inorganic fertilizers by half. Fish yields were the same with *Azolla* or inorganic fertilizers alone, whereas the yields of both fish and rice were higher in the combined *Azolla* and inorganic fertilizer regime (Table 2.5).

Shanmugasundaram and Ravi (1992) reported the use of *Azolla* (*A. microphylla*) as nitrogen fixing fertilizer and feed for Nile tilapia (*O. niloticus*) stocked in a low-lying ricefield (0.2 ha) in the Tanjore Deltaic Zone, Tamil Nadu, India. The ricefield was

TABLE 2.5
Use of *A. microphylla* as fertilizer in rice-fish culture system- fish species *(O. niloticus)*

Initial weight (g)	Fish density (Numbers/ha)	Duration (days)	Fertilizer regimes	Fertilizer rate (kg/ha)	Quantity of N (kg/ha)	Fish yield (kg/ha)	Rice yield (kg/ha)
8.9-9.4	5 000	75	Azolla only	3 750	5.63	45.1	2 567
			Inorganic fertilizer	150	38.5	45.0	3 096
			Azolla	3 750	5.6		
			Inorganic fertilizer	75	19.3	79.0	3 524

Source: Cagauan and Nerona (1986)

provided with trenches and connected to a main channel and the fish were raised in these trenches. The stocking density used was 6 000/ha for fingerlings weighing 19 g. Both fresh and dried *Azolla* were fed. Dried *Azolla* was incorporated in a supplemental fish feed and applied at 5 percent BW/day. The formula of this supplemental feed was stated to be *Azolla* (50 percent), rice bran (15 percent), chicken manure (10 percent), corn meal (5 percent), sorghum meal (5 percent), broken rice (2.5 percent) and groundnut cake (2.5 percent). The provision of water space for the fish lowers rice yields by about 300 kg/ha but the fish harvest compensates. Rice and fish culture yields a net income of US$258/crop/ha, compared to US$207/crop/ha for rice alone.

Furthermore, Shanmugasundaram and Balusamy (1993) reported the use of *Azolla* (*A. microphylla*) as feed to raise Indian major carps (catla, rohu and mrigal) stocked in low-lying wetlands in Bhavanisagar, Tamil Nadu, India. These authors used a 0.25 ha ricefield provided with trenches (1.0 m depth and width) to shelter the fish. Stocking density was 3 000/ha, using a 1:1:1 ratio of catla, rohu and mrigal. *Azolla* was applied twice at 2.0 tonnes/ha. Supplemental feed containing banana pseudostem and cow dung (1:1) was fed along with rice bran at 5 percent BW/ per day. Banana pseudostem and cow dung were incubated overnight before mixing with rice bran. Both rice and fish yields increased, with higher benefit cost ratios (1.88) in rice-fish-*Azolla*

TABLE 2.6
Economics of rice-fish-*Azolla* integration in India

Treatment	Rice yield (kg/ha)	Fish yield (kg/ha)	Gross return (US$/ha)	Net return (US$/ha)	Benefit cost ratio
Rice alone	8 765	-	822	353	1.75
Rice-fish	7 813	98.5	812	297	1.57
Rice-fish-*Azolla*	9 226	154.0	985	463	1.88

Source: Shanmugasundaram and Balusamy (1993)

TABLE 2.7
Results of rice-fish-*Azolla* integration highlighting increase in fish yield

Culture system	Fish species	Average harvest weight (g) With *Azolla*	Average harvest weight (g) Without *Azolla*	Yield (tonnes/ha) With *Azolla*	Yield (tonnes/ha) Without *Azolla*
Monoculture	*Oreochromis niloticus*	150-200	100-150	1.20	0.63
Polyculture	*Cyprinus carpio*	600	350	0.35	0.15
	Ctenopharyngodon idella	150	130	0.17	0.15
	Oreochromis niloticus	125	100	0.54	0.40
	Total			**1.06**	**0.70**

integration than rice-fish cultivation (1.57) (Table 2.6). Similarly, substantial increases in fish yield in rice-fish culture with *Azolla* compared to rice-fish without *Azolla* have been reported by Cagauan and Pullin (1991). Fish yields from rice-*Azolla*-fish culture trials were higher than those for rice-fish culture (Table 2.7). Yields from Nile tilapia in monoculture and from polyculture of common carp, grass carp and Nile tilapia were 1.20 and 1.06 tonnes/ha/year respectively from a rice-fish-*Azolla* system, compared with 0.63 and 0.70 tonnes/ha/year respectively from rice-fish fields without *Azolla*.

In rice farming systems, including rice-fish culture, *Azolla* is best incorporated as a fertilizer during its linear growth phase, when there is maximum productivity, low lignin content and therefore rapid decomposition. The value of *Azolla* as a fish feed is also highest during its linear growth phase. The crude protein content of *Azolla* is generally higher during this phase. The amino acid content of *Azolla* increases during the linear growth phase and falls sharply when the growth slows down with a corresponding increase in its lignin content. Digestibility clearly decreases after the linear growth phase with increasing lignin content (Van Hove *et al.*, 1987). It is therefore important to maintain an equilibrium between the population of fish and that of *Azolla*, either by introducing, when necessary, a supplementary biomass of *Azolla* collected elsewhere, or by harvesting the excess biomass in order to keep the *Azolla* population in the linear growth phase.

Pig-duck-fish-*Azolla* and fish-*Azolla* integration

Very few reports are available on the use of *Azolla* as fish feed in pond culture; however, there are reports of integrated studies. Majhi, Das and Mandal (2006) fed grass carp (*Ctenopharyngodon idella*) fingerlings with finely chopped *Azolla caroliniana* placed over a feeding basket under pond conditions. *Azolla* was well accepted by grass carp. The final weight gain of *Azolla*-fed fish was significantly higher compared to the control fish. The net profit for production of grass carp was US$0.12/m^2.

Gavina (1994) studied pig-duck-fish-*Azolla* integration. Nile tilapia were stocked in three earthen ponds with a uniform water depth of 50 cm. The ponds were fertilized with a mixture of dry pig and duck manure at the rate of 500 kg/ha. After initial manure application, the water level was increased to 80 cm in all three experimental ponds. The ponds were stocked at three densities: 10 000/ha; 20 000/ha; and 30 000/ha. All treatments were manured (pigs and ducks) with 100 kg fresh material/ha/day and supplemented with fresh *Azolla* at 200 g/m^2/week. The consumption of *Azolla* by fish was not monitored. However, it was observed that the fresh *Azolla* were seeded at a

TABLE 2.8
Weight gain comparisons of *Azolla*-fed fish

Fish	No. of fish	Initial weight (g)	Final weight (g)	Survival (percent)	Culture period (days)	Total weight increase (g)	SGR (percent)	*Azolla* feed coeff.
Grass carp[1]	30	54.7	118.7	100	112	1 920	0.69	49.0
Crucian carp[1]	30	75.0	110.8	100	112	1 074	0.35	31.2
Nile tilapia[1]	30	24.7	163.1	100	100	4 152	1.89	52.2
Silver carp[1]	39	96.8	92.2	76.7	112	-1 017	-0.04	0.0
Nile tilapia (15 percent inclusion)[2]	25	1.67	3.23	56	30	21.84	2.20	
Nile tilapia (45 percent inclusion)[2]	25	1.70	2.28	61.3	30	8.89	0.98	
Grass carp[3]	-	22.7	270.3	100	150		1.65	
Tilapia zillii[4]	15	2.2	4.7	93	91	34.88	0.83	

[1] modified from FAO (1988 cited by Cagauan and Pullin, 1991)
[2] Fiogbé, Micha and Van Hove (2004)
[3] Majhi, Das and Mandal (2006)
[4] Abdel-Tawwab (2008)

rate of 200 g/m^2/week (10 kg/50 m^2) and cleared by fish after 6 or 7 days. It was found that *Azolla* could be a viable source of supplementary feed, considering the high cost of commercial feeds. The study was conducted for a period of three months. Mean net yield varied between 8.22 and 10.97 kg/ha/day (3-4 tonnes/ha/year) at stocking densities ranging between 10-30 000/ha.

Weight gain comparisons of *Azolla*-fed fish were carried out by the Soil and Fertilizer Institute of the Hunan Academy of Agricultural Sciences (FAO, 1988 cited by Cagauan and Pullin, 1991) using grass carp, Nile tilapia, crucian carp (*Carassius auratus*) and silver carp (*Hypopthalmychthys molitrix*) (Table 2.8). The weight gain by *Azolla*-fed grass carp averaged 174 g/fish compared with 134 g/fish for Nile tilapia and 35.8 g/fish for crucian carp. A weight decrease of 4.6 g/fish was observed for silver carp.

3. Floating aquatic macrophytes – Duckweeds

Duckweeds are small (1-15 cm) free-floating aquatic plants with worldwide distribution. They are monocotyledons belonging to the family *Lemnaceae* (which is derived from the Greek word *'Limne'*, meaning pond) and are classified as higher plants or macrophytes, although they are often mistaken for algae and some taxonomists consider them as being members of the Araceae. Duckweeds serve as nutrient pumps, reduce eutrophication effects and provide oxygen from their photosynthesising activity. Duckweeds are often seen growing in thick blanket-like mats on still nutrient-rich fresh and slightly brackish waters. They do not survive in fast moving water (>0.3 m/sec) or water unsheltered from the wind. They grow at water temperatures between 6 and 33 °C (Leng, Stambolie and Bell, 1995).

3.1 CLASSIFICATION

Duckweed consists of four genera: *Lemna*, *Spirodela*, *Wolffia* and *Wolffiella*. So far, 37 species belonging to the four genera have been identified from different parts of the world. Selected species are listed in Table 3.1. Taxonomically the family is complicated by clonal characteristics (Culley *et al.*, 1981). The most commonly available species belong to the three genera *Lemna*, *Spirodela* and *Wolffia*. Illustrations of selected species of duckweeds are given in Figures 3.1 - 3.3. It is quite common for floating mats of duckweeds to consist of more than one species, e.g. *Lemna* and *Wolffia*.

Lemna is the largest genera of the family Lemnaceae. *Lemna* is among the most complex and confusing groups within the entire family. Landolt (1986) hypothesized that *Lemna disperna* and *Lemna gibba* are related as progenitor-derivative species and the former species differentiated from the latter one. Reduction of some structures such as frond size, number of nerves and the number of ovules in *Lemna disperna*, along with its narrower geographic distribution, support the hypothesis that it was derived from *Lemna gibba* or from a common ancestor. *Lemna disperna* has a chromosome number of 2n = 40, whereas the numbers 2n = 40, 50, 70 and 80 have been found in

TABLE 3.1
Classification of selected species of duckweeds

Lemna	Spirodela	Wolffia	Wolffiella
L. gibba	S. biperforata	W. arrhiza	W. caudate
L. disperna	S. intermedia	W. australiana	W. denticulata
L. gibba	S. oligorrhiza	W. columbiana	W. lingulata
L. japonica	S. polyrrhiza	W. microscopia	W. oblonga
L. minima	S. punctata	W. neglecta	W. rotunda
L. minor			
L. minuscula			
L. paucicostata			
L. perpusilla			
L. polyrrhiza			
L. turionifera			
L. trisulca			
L. valdiviana			

FIGURE 3.1
Spirodela sp.

Source: DWRP (1998)

FIGURE 3.2
Common duckweed, *Lemna minor* grown in a pond (Phu Tho Province, Viet Nam)

Courtesy of M.G. Kibria

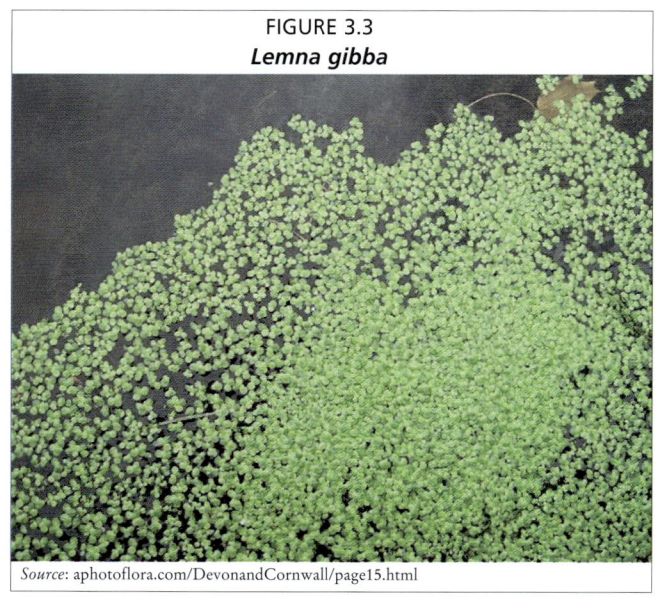

FIGURE 3.3
Lemna gibba

Source: aphotoflora.com/DevonandCornwall/page15.html

Lemna gibba (Crawford *et al.*, 2005). The allozyme study supports the continued recognition of two species and is concordant with the hypothesis that the species are related as progenitor and derivative. The reduced morphology of *Lemna disperna* and the allozyme data indicate that this species originated *via* dispersal of *Lemna gibba* or of a common ancestor of the two species.

3.2 CHARACTERISTICS

Duckweeds are adapted to a wide variety of geographic and climatic zones and are distributed throughout the world except in regions where temperature drops below 0 °C during part of the year. Most species are found in moderate climates of tropical and temperate zones. In deserts and extremely wet areas, duckweeds are rare. *Lemna* spp., for example are very rare in regions with high or very low precipitation and are not found in Greenland or the Aleutian Islands (Landolt, 2006). Although many species can survive extremes of temperature, they generally grow faster under warm and sunny conditions (Skillicorn, Spira and Journey, 1993). Most species show prolific growth in the tropics. Various microclimatic factors such as light intensity, salinity, regional temperature differences can influence the distribution of *Lemnaceae* species (Landolt, 1986). Birds and floods often disperse duckweeds to different geographic areas.

3.2.1 Reproduction

A duckweed plant consists of a single leaf or frond with one or more roots. Most species of duckweed multiply principally through vegetative propagation by the formation of daughter fronds from two pockets on each side of the narrow end of the frond (Gaigher and Short, 1986). Newly formed fronds remain attached to the mother frond during the initial growth phase and the plants therefore appear to consist of several fronds. Species of the genus *Spirodela* have the largest fronds, measuring as much as 20 mm across, while those *Wolffia* species are 2 mm or less in diameter. *Lemna* species are intermediate in size, being about 6-8 mm. An individual frond may produce as many as 20 daughter fronds during its lifetime, which lasts for a period of 10 days to several weeks. The daughter frond repeats the history of its mother frond. Some of the duckweed species, however, reproduce by producing unisexual and monoecious flowers and seeds. For example, *L. paucicostata* routinely flowers and seeds. However, the flowers are very small and rare in many species; male and female flowers are borne on the same plant. Each inflorescence generally consists of two male flowers and one female, but in *Wolffia*, there is one male and one female. The flowers are naked or surrounded by spathe. The fruit is a utricle and the seeds are smooth or ribbed. Vegetative reproduction is very rapid and is usually by the formation of buds of new fronds from the reproductive pouches (Guha, 1997).

Many species of duckweed survive at low temperatures by forming a special starchy 'survival' frond known as a turion. In cold weather, the turion is formed and sunk to the bottom of the pond where it remains dormant until warm water triggers resumption of normal growth. Several species survive at low temperatures without forming turions. During the winter season, the fronds are greatly reduced but remain at the surface. Occasionally, turion-like fronds will form, but the plants continue to slowly reproduce vegetatively. These plants are probably the best plants to utilize in a culture system, as restocking is virtually assured. *L. gibba*, *L. valdiviana*, *L. minor*, *L. trisulca* and *L. minuscula* are five duckweed species that frequently show some growth at cold temperatures.

3.2.2 Environmental requirements

A variety of environmental factors, such as water temperature, pH and nutrient concentration, control the growth and survivability of duckweeds. The other environmental factors that influence the growth rates of duckweed colonies are presence of toxins in the water, crowding by overgrowth of the colony and competition from other plants for light and nutrients. However, the growth rate of duckweed is favoured by organic pollutants as well as inorganic nutrients (Guha, 1997). The effect of these various factors is summarised below.

Temperature
Temperature tolerance and optima are dependent on species and possibly even on clones. Optimum temperature for maximum growth of most groups apparently lies between 17.5 and 30 °C (Culley *et al.*, 1981; Gaigher and Short, 1986). Although some species can tolerate near freezing temperatures, growth rate declines at low temperature. Below 17 °C some duckweeds show a decreasing rate of growth (Culley *et al.*, 1981). Most species seem to die if the water temperature rises above 35 °C. The effect of temperature on growth is affected by light intensity, i.e. as light increases, growth rates increase from 10 to 30 °C.

In Bangladesh, Khondker, Islam and Nahar (1993a) reported the temperature dependent growth of *S. polyrrhiza* with a maximum biomass of 76.4 g/m² recorded in the middle of February, after which the biomass depleted gradually with the rise in water temperature. The water temperature in the middle of February was about 19 °C. Islam and Khondker (1991) also obtained a high growth of *S. polyrrhiza* at a temperature of 16 °C. Furthermore, Khondker, Islam and Nahar (1993b) reported maximum growth of *S. polyrrhiza* at water temperatures of 22.2-22.5 °C in a growth study conducted in pond water. Khondker, Islam and Makhnun (1994) reported an inverse correlation between water temperature and the biomass of *L. perpusilla* when the water temperature varied between 15 and 28 °C. These authors also noted that the growth of this duckweed species ceased completely at 26 °C and above.

pH
Duckweeds are generally considered to have a wide range of tolerance for pH. They survive well from pH 5 to 9, although some authors put their range between 3 and 10. However, pH tolerance limits of the various species differ. Stephenson *et al.* (1980) noted that duckweed display optimum growth in a medium of pH 5.0-7.0. Generally, duckweeds grow best over the pH 6.5 to 7.5 range. A doubling of biomass in 2 to 4 days has been demonstrated at pH levels between 7 and 8 (Culley *et al.*, 1981). Unionized ammonia is the preferred nitrogen substrate for duckweed. An alkaline pH shifts the ammonium-ammonia balance toward the un-ionized state and results in the liberation of free ammonia, which is toxic to duckweed at high concentrations (100 mg NH_3/L).

Islam and Paul (1977) observed that *W. arrhiza* grew at a pH range of 5-10, although the optimum pH was found to be 7-8. In Bangladesh, *S. polyrrhiza* has been reported to grow best at a pH between 6.5 and 7.5 (Islam and Khondker, 1991). The range of pH for optimum growth of *S. polyrrhiza* reported in India was 6.8-8.5 (Kaul and Bakaya, 1976; Gopal and Chamanlal, 1991). Khondker, Islam and Makhnun (1994) reported the pH range of 6.9 and 7.8 to be suitable for the growth of *L. perpusilla*. Similarly, Van der Does and Klink (1991) observed pH of 7.36 in a lemnid habitat in the Netherlands supporting growth of *L. perpusilla*. A summary of minimum, maximum and optimum pH of various duckweed species is presented in Table 3.2.

Conductivity
Electrolyte conductivity appears to have some effect on the growth of different species of duckweed. Zutshi and Vass (1973) found *L. gibba* and *L. minor* growing in stagnant waters rich in electrolyte ranging from 400-500 µS/cm. Gopal and Chamanlal (1991) reported the maximum biomass of *L. perpusilla* and *S. polyrrhiza* from roadside pools and ditches in India within a electrolyte conductivity range of 650-1 000 µS/cm. Khondker, Islam and Nahar (1993a) recorded the complete disappearance of

TABLE 3.2
Minimum, maximum and optimum pH of various duckweed species

Duckweed species	Min	Max	Optimum	Reference
L. minor			6.1-6.7	Hicks (1932, cited by DWRP, 1997); McLay (1976)
L. perpusilla	3.2		6.9-7.8	Landolt and Kandeler (1987); Khondker, Islam and Makhnun (1994)
S. polyrrhiza	3.7		6.5-8.5	Gopal and Chamanlal (1991); Islam and Khondker (1991), Kaul and Bakaya (1976); Landolt and Kandeler (1987)
S. punctata			7.0	McLay (1976)
W. arrhiza	3.5	10.0	5-7.8	Hicks (1932, cited by DWRP, 1997); Islam and Paul (1977); Landolt and Kandeler (1987)
W. australiana			5.0	McLay (1976)
W. columbiana			6.4-7.0	Hicks (1932, cited by DWRP, 1997)

Source: DWRP (1998)

S. polyrrhiza by the end of May when a sharp fall in conductivity and alkalinity was observed. The electrolyte conductivity of water supporting the growth of *L. perpusilla* in Bangladesh reported by Islam and Khondker (1991) and Khondker, Islam and Makhnun (1994) were 625 µS/cm and 200-890 µS/cm, respectively. High electrolyte conductivity (1 090 µS/cm) of water supporting the growth of *L. perpusilla* was also reported by Van der Does and Klink (1991) in a lemnid habitat in the Netherlands.

Nitrogen

In general, temperature and sunlight control duckweed growth more than nutrient concentrations in the water. At high temperatures, duckweed can grow rapidly down to trace levels of phosphorus and nitrogen. The crude protein content of duckweed however, seems to increase to a maximum of ~40 percent DM over the range from trace ammonia concentrations to 7-12 mg N/L (Leng, Stambolie and Bell, 1995). Culley *et al.* (1981) reported that the TKN of water should not drop below 20-30 mg/l if the optimum production and a high crude protein content of duckweed are to be maintained.

Duckweeds prefer ammonia nitrogen (NH_4-N) as a source of nitrogen and will remove ammonia preferentially, even in the presence of relatively high nitrate concentrations. Lüönd (1980) demonstrated that higher growth rates were attained when nitrogen was in the NH_4-N rather than the NO_3-N form. In organically enriched waters, nitrogen tends to be concentrated in the NH_4-N rather than the NO_3-N form at pH levels below 9 and plant growth is equally efficient in anaerobic and aerobic waters (Said *et al.*, 1979). In lagoons receiving organic animal wastes, the pH seldom exceeds 8, particularly with a full duckweed cover that suppresses phytoplankton growth (Culley *et al.*, 1978). The plants can tolerate very high ionized ammonia (NH_4-N) concentrations but the effects of unionized ammonia (NH_3-N) have not been clearly demonstrated. Urea is a suitable fertilizer and is rapidly converted to ammonia under normal conditions. According to the results of laboratory experiments, duckweed tolerates concentrations of elemental N as high as 375 mg/l (Rejmánková, 1979).

Phosphorus and potassium

Phosphorus is essential for rapid growth and is a major limiting nutrient after nitrogen, although its quantitative requirement for maximum growth is generally low. Fast growing duckweed in nutrient rich water is a highly efficient sink for both phosphorus and potassium; little of each, however, is required for rapid growth. Saturation of phosphate uptake by duckweed occurs at available PO_4-P concentrations of 4 to 8 mg/l. Duckweed growth is not particularly sensitive to potassium or phosphorus once an adequate threshold has been reached. Rejmánková (1979) reported good growth of duckweed within the P concentrations of 6 to 154 mg/l. Culley *et al.* (1978), working in dairy waste lagoons, achieved doubled production from 2 to 4 days at P concentrations in excess of 35 mg/l. Reduced growth in some species occurs only after P values dropped below 0.017 mg/l (Lüönd, 1980). Khondker, Islam and Makhnun (1994) observed that both phosphate and silicate concentrations had significant positive correlation with the biomass of *L. perpusilla* in Bangladesh.

Other minerals
A range of other important mineral levels found in water supporting Lemnaceae is presented in Table 3.3. Although these minerals are essential for their survival, duckweed growth is not particularly sensitive to potassium or phosphorus once an adequate threshold has been reached.

TABLE 3.3
Range of important mineral contents (mg/l) of water supporting Lemnaceae

Parameter	Absolute range	Range of 95 percent of the samples
K	0.5 – 100	1.0 – 30
Ca	0.1 – 365	1.0 – 80
Mg	0.1 – 230	0.5 – 50
Na	1.3 – >1 000	2.5 – 300
HCO_3	8 – 500	10.0 – 200
Cl	0.1 – 4 650	1.0 – 2 000
S	0.03 – 350	1.0 – 200

Source: modified from Landolt (1986)

In summary
Maximum, minimum and optimum requirements of some of the most important environmental parameters (temperature, pH, conductivity, nitrogen and phosphorus) are given in Table 3.4. It is apparent that the duckweeds are robust in terms of survival, but sensitive in terms of thriving. They have extreme range of tolerance for temperature, pH, conductivity, nitrogen and phosphorus with well-defined range of optimum requirement.

3.3 PRODUCTION
3.3.1 Background information
Duckweed growth is largely a function of available nutrients, temperature, light, and degree of crowding. The highest growth rate reported for Lemnaceae under optimal laboratory conditions is about 0.66 generations per day, which corresponds to a doubling time of 16 hours (DWRP, 1997). Duckweeds generally double their mass in 16 hours to 2 days under optimal nutrient availability, sunlight, and water temperature. An individual plant, a small leaflet (frond), produces 10 to 20 daughter fronds during its lifetime, which lasts for a period of 10 days to several weeks. The daughter frond repeats the history of its mother frond. This results in an exponential growth, at least until the plants become crowded or run out of nutrients. Frequent periodic removal of the plants encourages continuation of the exponential growth.

TABLE 3.4
Summary of some important environmental requirements of duckweed

Environmental parameters	Minimum	Maximum	Optimum
Temperature (°C)	>0	35	15-30
pH	3.0	10.0	6.5-8.0
Conductivity (µS/cm)[1]	200	1 090	n.s.
Nitrogen (mg/l NH_3-N)	Trace	375	7-12
Phosphorus (mg/l PO_4-P)	0.017	154	4-8

[1] Conductivity range found supporting growth of duckweed.

The individual clones of the same species may show quantitative variation in growth characteristics (Rejmánková, 1975; Porath, Hepher and Koton, 1979). In the Czech Republic, Rejmánková (1975, 1979) reported maximum dry matter yields of 3.14-3.54 g and 7.09 g/m²/day from unmanaged fish ponds and outdoor tanks respectively, when weekly harvesting was done. Rejmánková (1981) further reported that an estimated annual net dry matter production of 7.5-8.0 tonnes/ha could be obtained, provided nutrients and crowding were not limiting and harvesting was frequent.

Culley and Myers (1980) and Said *et al.* (1979) working in the southern USA (9-10 months growing season) demonstrated that high nutrient lagoons and outdoor tanks (enriched with cattle manure) yielded the dry matter equivalent of about 15 g/m²/day (55 tonnes/ha/year) when regular daily harvesting was done to remove the excess. Said *et al.* (1979) reported an annual dry weight yield of 44 tonnes/ha or about 12 g/m²/day. Furthermore, Culley and Myers (1980) obtained an estimated average annual dry matter production of 23.3 tonnes/ha with daily harvesting ranging from 10 to 35 percent of the standing crop, depending on the season. In a sewage-fed culture system, the growth rate of *Azolla* spp., *Spirodela* spp. and *Wolffia* sp. were found to be 160, 350 and 280 g/m³/day, respectively (Reddy *et al.*, 2005).

Table 3.5 presents the yields of various duckweed species under different environmental conditions. The values varied widely, ranging from 9 to 38 tonnes (DM)/ha/year. This wide range of productivity may be attributed to differences in species, climatic conditions, nutrient supply and environmental conditions. Many of the reported high yields are based on extrapolated data obtained from short-term growth from small-scale experimental systems rather than potential long-term yields from commercial-sized systems. Edwards (1990) reported extrapolated yields of ~20 tonnes (DM)/ha/year of *Spirodela* from experiments that were carried out for periods of 1-3 months in septage-fed 200 m² ponds in Thailand; however, the yield declined to the equivalent of ~9 tonnes (DM)/ha/year over a 6 months period. Based on the available data and the foregoing discussion, it may therefore be concluded that an average annual yield of around 10-20 tonnes/DM/ha can be obtained from an aquatic environment where nutrients are generally not limiting and frequent harvesting is practised to avoid plant overcrowding.

3.3.2 Duckweed farming

Duckweed farming is a continuous process requiring intensive management for optimum production. Daily attention and frequent harvesting are needed throughout the year to ensure optimum productivity. Duckweed can grow in water of any depth.

TABLE 3.5
Yields of various duckweed species under different environmental conditions

Species	Environmental condition	Yield (dry matter tonnes/ha/year)	Reference
L. minor	UASB effluent	10.7	Vroon and Weller (1995)
L. minor	Nutrient non-limiting water	16.1	Reddy and DeBusk (1984)
L. perpusilla	Septage-fed pond	11.2	Edwards, Pacharaprakiti and Yomjinda (1990)
L. perpusilla, S. polyrrhiza and *W. arrhiza*	Septage from septic tank	9.2-21.4	Edwards *et al.* (1992)
Lemna	Domestic wastewater	26.9	Zirschky and Reed (1988)
Lemna	Sugar mill effluent	32.1	Ogburn and Ogburn (1994)
Lemna, Spirodela and *Wolffia*	Domestic wastewater	13-38	Skillicorn, Spira and Journey (1993)
Lemna and *Wolffia*	Faecally polluted surface water	14-16	Edwards (1987)
S. polyrrhiza	Domestic wastewater	17-32	Alaerts, Mahbubar and Kelderman (1996)
S. polyrrhiza	Sewage effluent	14.6	Sutton and Ornes (1975)
S. polyrrhiza	Nutrient non-limiting water	11.3	Reddy and DeBusk (1985)

It will grow in as little as one centimetre of water. A pond depth of between 20 and 50 cm is generally recommended to reduce the potential sources of stress and to facilitate harvesting (Gaigher and Short, 1986). Duckweeds are prone to be blown into heaps by heavy winds or wave action. This allows light to penetrate the water column and would stimulate phytoplankton and algal growth. If the plants become piled up in deep layers, however, the lowest layer will be cut off from light and will eventually die (Skillicorn, Spira and Journey, 1993). Plants pushed from the water onto a bank will also dry out and die. Long narrow ponds that are sited perpendicular to the common wind are recommended. Dividing the pond into smaller segments by using bamboo can also mitigate the adverse effects of wind. An NGO called 'PRISM' applied a grid of bamboo poles (Figure 3.4) of approximately 5 x 5 m in large ponds and wide canals. This functioned satisfactorily for all conditions met up to that date in Bangladesh (DWRP, 1998). Lemma USA Inc. promotes floating barrier grids made of polyethylene that will reduce wind and wave action for its wastewater treatment plants. The sides of the ponds must preferably be vertical to prevent the plants from becoming stranded and at least 10 cm higher than the water level to accommodate heavy rains. The ponds must be fed with effluent through furrows rather than pipes because the latter tend to become clogged. Several inlets must be provided to spread the inflowing nutrients over the pond.

Since the growth of duckweed is dependent on water temperature, pH and nutrient concentration, these factors need to be balanced and maintained within reasonable limits for duckweed to thrive. The management strategies for duckweed culture should therefore focus on when to fertilize, harvest, and buffer; how much to fertilize and to harvest; and which nutrients to supply. Appropriate management should be aimed at maintaining a complete and dense cover of duckweed, low dissolved oxygen, and a pH of 6-8. A dense cover shuts out light and suppresses the growth of algae, which minimizes CO_2 production from algal respiration and prevents its elevating effect on pH.

Any waste organic material that is readily biodegradable and has a sufficiently high nutrient content could be used for duckweed cultivation. The most economical sources of such waste materials are all kinds of animal manure, kitchen wastes, wastes from a wide range of food processing plants, biogas effluents, and slaughterhouse wastes. Solid materials, such as manure from livestock, night soil from villages, or food processing wastes, can also be mixed with water and added to ponds at suitable levels. All wastewater containing manure or night soil must undergo an initial treatment by holding it for a few days in an anaerobic pond, before using it to cultivate duckweed.

Sutton and Ornes (1975) and Said *et al.* (1979) demonstrated the necessity of periodic additions of nutrients to small duckweed culture systems receiving municipal or dairy cattle wastes. Within 1-3 weeks, there was a noticeable drop in N, P and K within the plants. There was a corresponding drop in crude protein as the plant nitrogen declined. In unmanaged ponds, where duckweeds are not routinely harvested, the plants quickly become crowded and those beneath the surface die back.

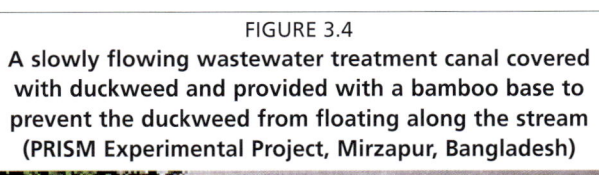

FIGURE 3.4
A slowly flowing wastewater treatment canal covered with duckweed and provided with a bamboo base to prevent the duckweed from floating along the stream (PRISM Experimental Project, Mirzapur, Bangladesh)

Source: DWRP (1998)

Due to the high nitrogen requirement of duckweed and the relatively rapid loss of nitrogen from aquatic system, this nutrient tends to be limiting in ponds fed with wastewater (Gaigher and Short, 1986). Studies at Louisiana State University have shown that the nitrogen conversion efficiency from agricultural waste to duckweed is only about 30 percent under normal field conditions (Culley et al., 1981). Large-scale duckweed production therefore requires the availability of relatively large quantities of organic waste. The addition of cheap inorganic nitrogen could also therefore improve the wastewater conversion efficiency. The other nutrients that are needed for optimum growth of duckweed are phosphorus, potassium and trace minerals.

Fertilization

Urea is a suitable fertilizer, containing approximately 45 percent nitrogen, and is rapidly converted to ammonia under normal conditions. Muriate of potash (MP) and triple superphosphate (TSP) are commercial sources of potassium and phosphorus that are widely available in most countries and have been used where duckweeds have been farmed. Duckweed growth is not particularly sensitive to potassium or phosphorus once an adequate threshold has been reached. A ratio of TSP to urea of 1:5 worked satisfactorily in an experimental duckweed production programme in Bangladesh (Skillicorn, Spira and Journey 1993). Similarly, a ratio for MP to urea of 1:5 was found to be satisfactory for good production in the same duckweed production programme in Bangladesh.

Nutrients are absorbed through all surfaces of the duckweed leaf (Leng, Stambolie and Bell, 1995). There are at least three methods of fertilizer application including broadcasting, dissolving in the water column of the plot, and spraying a fertilizer solution on the duckweed mat.

A fertilizer application matrix aiming to achieve variable daily production ranging from 500-1 000 kg of fresh duckweed per hectare was developed by PRISM in their experimental programme at Mirzapur, Bangladesh (Table 3.6). Furthermore, PRISM recommended daily fertilization rates for different types of duckweed (Table 3.7). The application rate varies from 21-28 kg/ha/day (amounting to >7 tonnes/ha/year) with an anticipated fresh biomass yield of 900-1 000 kg/ha/day. The daily fertilization rate for duckweed cultivation developed by the Bangladesh Fisheries Research Institute (BFRI) is presented in Table 3.8. The fertilizer schedules developed by PRISM and BFRI are very similar (Tables 3.7 and 3.8), except that BFRI recommended half the dosage of inorganic fertilizer when cow dung was used at the rate of 750 kg/ha/year.

TABLE 3.6

Dayly fertilizer application matrix for duck weed cultivation developed by PRISM in their experimental programme at Mirzapur, Bangladesh

Fertilizer application (kg/ha)	500	600	700	800	900	1 000
	Daily production of fresh plants (kg/ha)					
Urea	10.0	12.0	14.0	16.0	18.0	20.0
TSP	2.0	2.4	2.8	3.2	3.6	4.0
MP	2.0	2.4	2.8	3.2	3.6	4.0
Sea salt	4.5	5.4	6.3	7.2	8.1	9.0

Source: Skillicorn, Spira and Journey, (1993)

Seeding

Seeding is a highly important management measure since a full duckweed cover should be established before any algal bloom can start dominating the water body. The seed rate advised is 60 kg/100 m^2 for *Spirodela* spp. and *Wolffia* spp. and 40 kg/100 m^2 for *Lemna* spp. in order to obtain a dense cover in 3 days time (DWRP, 1998). From day four onwards daily harvesting can start.

Stress management

Stress management of the crop is necessary particularly during very hot and dry weather. 'Dunking' (dipping the duckweed below the water surface) once a day as a regular crop maintenance practice is recommended; this reduces the stress from overheating. Dunking consists of agitating the whole-cultivated area by hand until all plants have been physically immersed and wetted.

Plant density and harvesting rate

The productivity of duckweed increases with increasing plant density up to a density where the plants completely cover the surface of the water, and then remains constant. In order to maintain good productivity and prevent competition by phytoplankton/suspended algae, the density must be maintained at this level or a slightly higher level. Competition between phytoplankton/suspended algae and duckweed is a potential constraint to the cultivation of the latter in nutrient-rich water. Phytoplankton smothers the roots of duckweed, which then turn yellow in colour, suffer a decline in growth rate, and eventually die. The development of an algal bloom can also reduce nutrient availability and thus eventually reduce the growth of duckweed.

An optimum standing crop density is a cover that is complete but which still provides enough space to accommodate rapid growth of the colony. In the PRISM experimental programme at Mirzapur, Bangladesh a base *Spirodela* density of 600 g/m^2 was shown to yield a daily incremental growth of 50 to 150 g/m^2/day (Skillicorn, Spira and Journey, 1993). This is equivalent to a daily fresh (wet weight) crop production rate of 0.5 to 1.5 tonnes/ha. These authors recommended a plant density of 400 to 800 g/m^2 for optimum production. BFRI (1997) obtained duckweed production of 700-1 500 kg/ha/day at plant densities varying from 400-600 g/m^2 in their experimental programme at Mymensingh.

High-density populations contain a high ratio of old fronds, which can be detrimental in various ways. Duckweed should therefore be harvested frequently, preferably daily. The standing crop density, or the weight of fresh plant per square meter, will determine the amount and timing of harvests. Daily harvesting of the incremental growth of the duckweed plot - averaging approximately 100 g/m^2/day is recommended (Skillicorn, Spira and Journey, 1993). Culley and Myers (1980) obtained an annual dry weight production of 23.31 tonnes/ha with daily harvesting ranging from 10 to 35 percent of the standing crop each day, depending on the season. Edwards (1990) recommended 25 percent harvesting of the duckweed biomass when duckweed growth completely covers the pond, with the remaining 75 percent left in the pond for further growth.

TABLE 3.7

Rates of fertilization application for duckweed cultivation techniques developed by PRISM

Duckweed	Rate of application (kg/ha/day)		
	Urea	TSP	MP
Spirodela	20	4	4
Wolffia	15	3	4
Lemna	15	3	3

Source: DWRP (1998)

TABLE 3.8

Rates of fertilization application for duckweed cultivation techniques developed by Bangladesh Fisheries Research Institute (BFRI)

Fertilizer combination	Rate of application (kg/ha/day)			
	Urea	TSP	MP	Cow dung
Inorganic fertilizer only	15-20	3-4	3-4	-
Combination of organic and Inorganic fertilizer	7.5	1.5	1.5	750

Source: BFRI (1997)

This author opined that this harvest could be made every 1-3 days, depending on the season.

Duckweed in wastewater treatment

Ferdoushi *et al.* (2008) tested the efficacy of *Lemna* and *Azolla* as biofilters of nitrogen and phosphate in fish ponds in Bangladesh and found that they removed the excess amount of nutrients from the water body and maintained sustainable environmental conditions. Duckweeds have received much attention because of their potential to remove contaminants from wastewater (Leng, Stambolie and Bell, 1995). Duckweed wastewater treatment systems have been studied for dairy waste lagoons (Culley *et al.*, 1981), raw domestic sewage (Oron, 1994; Skillicorn, Spira and Journey, 1993; Alaerts, Mahbubar and Kelderman, 1996), secondary effluent (Harvey and Fox, 1973), waste stabilization ponds (Wolverton, 1979) and fish culture systems (Porath and Pollock, 1982; Rakocy and Allison, 1981). The basic concept of a duckweed wastewater treatment system is to farm local duckweed on the wastewater requiring treatment. Duckweed has a high mineral absorption capacity and can tolerate high organic loading as well as high concentrations of micronutrients.

Duckweed wastewater treatment systems remove, by bioaccumulation, as much as 99 percent of the nutrients and dissolved solids contained in wastewater (Skillicorn, Spira and Journey, 1993). These substances are then removed permanently from the effluent stream following the harvesting of a proportion of the crop. The plants also reduce suspended solids and BOD by reduction of sunlight in lagoons. Duckweed systems distinguish themselves from other effluent wastewater treatment mechanisms in that they also produce a valuable, protein-rich biomass as a by-product.

Depending on the wastewater, the harvested crop may serve as an animal feed, a feed supplement supplying protein/energy and minerals, or a fertilizer. The question of toxic elements must be considered if certain types of waste material serve as the nutrient source for duckweed culture; for example, duckweed will absorb heavy metals and insecticides from the wastewater. It may, therefore, have to be decontaminated prior to feeding to animals if heavy metals are present in the water.

Landolt and Kandeler (1987) reported that of all aquatic plants, Lemnaceae have the greatest capacity in assimilating the macro-elements N, P, K, Ca, Na and Mg. Table 3.9 presents some data on daily nitrogen and phosphorus uptake efficiency by duckweed. The results from the various studies are not comparable because different species are used and different climatological and operational conditions were applied. Temperature may have a significant effect on nutrient uptake efficiency as has also been observed for other aquatic plants.

TABLE 3.9
Daily nitrogen and phosphorus uptake by duckweed

Region/Country	Species	Uptake (g/m^2)	
		N	P
Italy	*L. gibba/ L. minor*	0.42	0.01
CSSR	Duckweed	0.20	-
USA	*Lemna* sp.	1.67	0.22
Louisiana, USA	Duckweed	0.47	0.16
India	*Lemna* sp.	0.50-0.59	0.14-0.30
Minnesota, USA	*Lemna* sp.	0.27	0.04
Florida, USA	*S. polyrrhiza*	-	0.015

Source: adapted from DWRP (1997)

Culley *et al.* (1981) made a comprehensive study on nutrient uptake from wastewater by a mixed culture of duckweed (Table 3.10). This shows that duckweeds are capable of removing considerable amounts of organic wastes from natural water. An annual nutrient removal capacity covered by *Lemnaceae* of 1 378 kg TKN, 345 kg P, and 441 kg K per hectare of water area was calculated.

Summarizing the results of PRISM Experimental Site at Mirzapur, Skillicorn, Spira and Journey (1993) reported that treating an average flow of 125 m^3/day of hospital, school, and residential wastewater produced by a population of between 2 000 and 3 000 persons, the 0.6 ha duckweed treatment plant produces a final treated effluent that exceeds the highest quality standards mandated in the USA (Table 3.11). These authors also estimated that a typical duckweed wastewater treatment plant would yield a daily harvest of up to one ton of duckweed plants (wet weight) per hectare or 90 kg per hectare of dried, high protein duckweed meal each day.

TABLE 3.10

Mean annual dry duckweed yield and nutrient uptake by a mixed culture of duckweed[1] harvested daily[2] from a 25 m² lagoons[3] receiving dairy cattle wastes[4]

	Duckweed yield (kg/m²)	Duckweed[5]		
		TKN (g/m²)	P (g/m²)	K (g/m²)
Dec-Feb	0.195	11.5	2.9	3.7
Mar-May	0.576	34.2	8.5	10.9
Jun-Aug	1.020	60.2	15.1	19.3
Sep-Nov	0.540	31.9	8.0	10.2
Total (kg/ha)	23 310	1 378	345	441

[1] *S. polyrrhiza, S. punctata, L. gibba* and *W. columbiana* in approximate equal amounts at Baton Rouge, Louisiana, USA (9-10 months growing season)
[2] September-February: a mean of 105 of duckweed removed daily; March-August: a mean of 35 percent of duckweed removed daily
[3] Trials were run in triplicate
[4] Fresh manure loading in first stage lagoon provided an effluent to the test lagoons with 15-65 mg/l TKN; 18-28 mg/l phosphorus (P); 38-69 potassium (K); pH 7.6-7.9
[5] TKN 5.9 percent of dry weight, P 1.48 percent and K 1.89 percent

Source: modified from Culley *et al.* (1981)

TABLE 3.11

Quality of final treated effluent at Mirzapur Experimental Site on 23 March 1991

Treatment phase	BOD$_5$ (mg/l)	NH$_3$ (mg/l)	P (mg/l)	Turbidity (FTU[1])
Raw influent	120	39.40	1.90	113
Primary	60	32.20	2.00	85
Duckweed	1	0.03	0.03	10
US Summer Standards: Washington D.C. area	10	2.00	1.00	20

[1] This turbidity unit standard is roughly equivalent to total suspended solids (TSS) times two

Source: Skillicorn, Spira and Journey (1993)

3.4 CHEMICAL COMPOSITION

Each frond of duckweed absorbs nutrients through the whole plant, not through a central root system, directly assimilating organic molecules such as simple carbohydrates and various amino acids. The entire body is composed of non-structural, metabolically active tissue; most photosynthesis is devoted to the production of protein and nucleic acids, making duckweeds very high in nutritional value. The nutritional content of duckweed is probably more dependent on the mineral concentrations of the growth medium than on the species or their geographic location. Water low in nutrients generally results in reduced nutritional content in duckweed. Crude fibre content is generally lower (varying between 7-10 percent) for duckweed grown in nutrient-rich water than that grown in nutrient-poor water (11-17 percent).

Compared with most plants, duckweed leaves have little fibre (5 percent in cultured plants) as they do not need to support upright structures (Leng, Stambolie and Bell, 1995). Crude fibre content was generally lower, varying between 7-10 percent, for duckweed grown in nutrient-rich water as opposed to 11-17 percent for duckweed grown in nutrient-poor water. In general, the ash content ranges between 12-18 percent (Leng, Stambolie and Bell, 1995).

Duckweeds are known to accumulate large amounts of minerals in their tissues. Skillicorn, Spira and Journey (1993) reported that fibre and ash contents are higher and protein content lower in duckweed colonies with slow growth. Duckweeds are rich source of nitrogen, phosphorous, potassium and calcium (Guha, 1997). The concentration of N and P in duckweed tissues depend on the amount of N and P in the water, up to a threshold concentration that has not been clearly defined. Above this threshold, there is little increase in the tissue. Culley *et al.* (1978) suggested that under lagoon conditions, 20-30 mg/l TKN might be required to maintain a crude protein level above 30 percent. The crude protein content of duckweeds grown on various nutrient solutions ranges from 7 to 45 percent of the plant dry weight, depending on the nitrogen availability (Culley *et al.*, 1981). When conditions are good, duckweed contains considerable protein, fat, starch and minerals, which appear to be mobilized for biomass growth when nutrient concentrations fall below the critical levels for growth. Nutrient contents in duckweed may therefore vary according to the conditions in which it is grown. Slow growth, starvation and aging have been reported to result in protein levels as low as 7 percent DM (Landolt and Kandeler, 1987).

A summary of the nutritional composition of different species grown under different environmental conditions is presented in Table 3.12. Fresh duckweed contained about 91-95 percent water and the moisture content is apparently not influenced by the medium under which it was grown. Duckweed species grown under nutrient-poor water or under sub-optimum nutrient conditions have crude protein contents varying between 9-20 percent, while the level varied from 24-41 percent for duckweed species grown in nutrient-rich water. The crude protein content of duckweed seems to increase from trace ammonia concentrations to 7-12 mg N/L when crude protein reaches a maximum of about 40 percent (Leng, Stambolie and Bell, 1995). Similarly, the lipid content was lower (1.8-2.5 percent) in duckweed species grown in nutrient-poor water, while it generally varied between 3-7 percent for duckweed grown in nutrient-rich water. The medium in which duckweed was grown or the nutrient status of water did not influence the ash content of duckweed (Leng, Stambolie and Bell, 1995). Skillicorn, Spira and Journey (1993) reported that fibre and ash contents are higher and protein content lower in duckweed colonies with slow growth.

Studies by Porath, Hepher and Koton (1979) and Rusoff, Blakeney and Culley (1980) show clearly that the duckweed indeed has high quality protein. It has a better essential amino acid profile than most plant proteins and more closely resembles animal protein than any other plant proteins. According to Guha (1997), the protein of duckweeds is rich in certain amino acids that are often rather low in plant proteins. The nutritional value of Lemnaceae can be compared favourably with that of alfalfa in terms of lysine and arginine, two amino acids important in animal feeds. Duckweeds are rich in leucine, threonine, valine, isoleucine and phenylalanine and are low in methionine and tyrosine.[1] Some information on the amino acid content of various aquatic macrophytes is contained in Annex 1. Annex 1 Table 3 shows mean values determined for amino acids in four species of duckweed. It is evident that the values for the essential amino acids compare favourably with the FAO reference pattern, with the exception of methionine. The levels of amino acids are very similar in the various species and all the essential amino acids were generally present.

Cultured duckweed has high concentrations of trace minerals and pigments, especially β-carotene and xanthophyll (Haustein *et al.*, 1988). Duckweeds store varying amounts of calcium as calcium oxalate crystals in the vacuoles. Calcium oxalate may be toxic in large doses and the amount should be reduced to make duckweeds more

---

[1] www.mobot.org/jwcross/duckweed/nutritional-composition.htm

TABLE 3.12
Chemical analyses of various duckweed species grown under different environmental conditions

Duckweed species	Aquatic environment	Moisture (percent)	Proximate composition[1] (percent DM)					Minerals[1] (percent DM)		Reference
			CP	EE	Ash	CF	NFE	Ca	P	
L. gibba, USA	Low nutrient lagoon[2]	n.s.	9.4	1.8	16.8	17.0	55.5[6]	1.38	0.72	Culley et al. (1981)
L. minor, Bangladesh	Pond, nutrient status not specified	92.0	14.0	1.9	12.1	11.1	60.9	n.s.	n.s.	Zaher et al. (1995)
L. minor, Bangladesh	Ditch, nutrient status not specified	93.8	20.3-23.5	n.s.	n.s.	n.s.	n.s.	n.s.	n.s.	Majid et al. (1992)
L. polyrrhiza, India, raw leaf meal	Freshwater	32.5	18.6	1.5	2.5	11.0	66.4[6]	n.s.	n.s.	
L. polyrrhiza, fermented, India	...	n.s.	11.4	1.0	n.s.	7.5	n.s.	n.s.	n.s.	
S. polyrrhiza, USA	Low nutrient lagoon[2]	n.s.	13.1	2.5	13.3	16.1	55.0[6]	1.21	0.56	Culley et al. (1981)
S. polyrrhiza, Bangladesh	Ditch, nutrient status not specified	95.0	17.3-28.4	n.s.	n.s.	n.s.	n.s.	n.s.	n.s.	Majid et al. (1992)
S. punctata, USA	Low nutrient lagoon[2]	n.s.	10.6	2.3	14.1	11.3	61.7[6]	0.98	0.61	Culley et al. (1981)
W. arrhiza, whole plant, Bangladesh	Ditch, nutrient status not specified	91.2	14.9	n.s.	n.s.	n.s.	n.s.	n.s.	n.s.	Majid et al. (1992)
L. gibba, USA	High nutrient lagoon[3]	n.s.	36.3	6.3	15.5	10.1	31.8[6]	1.81	2.60	Culley et al. (1981)
L. gibba, USA	Dairy cattle waste lagoon	n.s.	38.5	3.0	16.4	9.4	32.7[6]	1.00	1.60	Hillman and Culley (1978)
L. minima, USA	Source not specified	n.s.	31.0	2.0	14.0	10.0	42.0	n.s.	n.s.	Shireman, Colle and Rottmann (1977)
L. perpusilla, Thailand	Septage-fed earthen pond	94.0-94.3	25.3-29.3	3.8-4.5	15.4-17.6	6.9-7.6	n.s.	n.s.	n.s.	Hassan and Edwards 1992
S. oligorrhiza, USA	Dairy cattle waste lagoon	n.s.	37.8	3.8	12.0	7.3	39.1[6]	1.30	1.50	Hillman and Culley (1978)
S. oligorrhiza, USA[4]	Treated wastewater effluent	n.s.	32.7	6.3	20.3	13.5	27.2[6]	1.49	1.15	Culley and Epps (1973)
S. oligorrhiza, USA[4]	Untreated septic tank influent	n.s.	32.3	n.s.	n.s.	n.s.	n.s.	1.29	1.17	Culley and Epps (1973)
S. oligorrhiza, USA[4,5]	Anaerobic swine waste lagoon	n.s.	41.4	5.1	12.9	8.3	32.3[6]	0.91	2.07	Culley and Epps (1973)
S. polyrrhiza, Thailand	Septage-fed earthen pond	91.0	23.8	3.8	18.3	11.7	42.4[6]	n.s.	n.s.	Hassan and Edwards (1992)
S. polyrrhiza, USA	High nutrient lagoon[3]	n.s.	39.7	5.3	12.8	9.3	32.9[6]	1.28	2.10	Culley et al. (1981)
S. polyrrhiza, USA	Dairy cattle waste lagoon	n.s.	40.9	6.7	12.9	8.7	30.8[6]	2.10	1.40	Hillman and Culley (1978)
S. punctata, USA	High nutrient lagoon[3]	n.s.	36.8	4.8	15.2	9.7	33.5[6]	1.75	1.50	Culley et al. (1981)

[1] CP = crude protein; EE = ether extract; CF = crude fibre; NFE = nitrogen free extract; Ca = calcium; P = phosphorus
[2] Low nutrient lagoon contained less than 5 mg/l TKN
[3] High nutrient lagoon contained more than 30 mg/l TKN
[4] Proximate composition and mineral content values corrected to dry basis
[5] Mean of eight values sampled over five months period
[6] Adjusted or calculated; not as cited in original publication

nutritious and digestible (Franceschi, 1989). The metabolic precursor of oxalate is L-ascorbic acid (vitamin C). A study with water lettuce (*Pistia stratiotes*) (see section 5) indicates that L-ascorbate and oxalate are synthesized within the crystal idioblast cells (Kostman *et al.*, 2001).

3.5 USE AS AQUAFEED

Because of its attractive nutritional qualities and the relative ease of production, a significant number of studies have been carried on the potential utilization of duckweed biomass as fish feed (Shireman, Colle and Rottmann 1977, 1978; Hillman and Culley, 1978; Stephensen *et al.*, 1980; Gaigher, Porath and Granoth, 1984; Naskar *et al.*, 1986; Hassan and Edwards, 1992). Available literature indicates that duckweeds are fed to fish in fresh form as a sole feed or in combination with other feed ingredients (Figure 3.5). Duckweeds are also fed as a dried meal ingredient in pelleted diets. Intensive fish production with duckweed as a predominant feed constituent has been reported by a number of authors (Hepher and Pruginin, 1979; Robinette, Brunson and Day, 1980; Culley *et al.*, 1981; Landolt and Kandeler, 1987, Skillicorn, Spira and Journey, 1993). Research studies on the use of duckweed as fish feed have been carried out under laboratory as well as under field conditions. Successful results have also been obtained on the on-farm utilization of duckweed as fish.

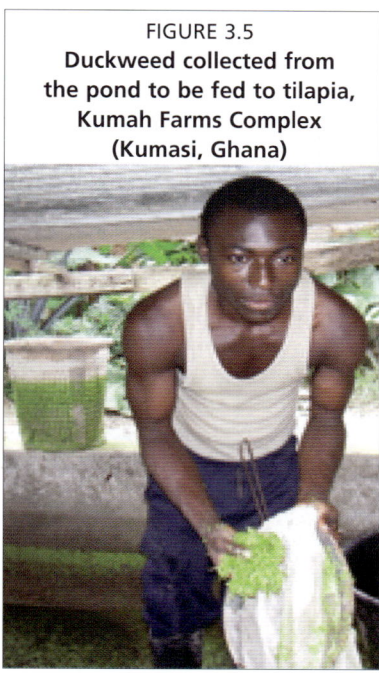

FIGURE 3.5
Duckweed collected from the pond to be fed to tilapia, Kumah Farms Complex (Kumasi, Ghana)

3.5.1 Laboratory studies

Successful feeding trials for grass carp with duckweed have been carried out since the early 1960s. Studies on the consumption of duckweed by aquatic animals have generally been confined to this species, although more recently feeding trials have also been carried out with others, including common carp, catfish, Indian major carps and tilapia.

Results on the use of duckweed as a feed for grass carp are generally very positive (Galkina, Abdullaev and Zacharova, 1965; Nikolskij and Verigin, 1966; Fischer, 1968, 1970; Edwards, 1974; Porath and Koton, 1977; Shireman, Colle and Rottmann, 1977, 1978; Baur and Buck, 1980; Hajra and Tripathi, 1985). Galkina Abdullaev and Zacharova (1965) reported that the grass carp showed more rapid growth when using duckweed than other feed materials. Porath and Koton (1977) noted that the weight of grass carp could be tripled (from 100 g to 300 g) within 50 days when feeding a mixture of *L. gibba* and *L. minor*.

Fresh duckweeds have also been efficiently utilized by common carp, catfish, Indian major carps and tilapia (Hepher and Pruginin, 1979; Robinette, Brunson and Day, 1980; Stephensen *et al.*, 1980; Gaigher, Porath and Granoth, 1984; Naskar *et al.*, 1986; Hassan and Edwards, 1992).

Summary results of selected growth studies carried out on the use of fresh and dried duckweed as feed for different fish species are presented in Tables 3.13 and 3.14. Fresh and dried duckweed were fed to grass carp, Nile tilapia, common carp, Indian major carps (rohu and mrigal), silver carp, Java barb, hybrid grass carp and hybrid tilapia. The duckweed species evaluated were *L. gibba*, *L. perpusilla*, *L. minima*, *L. minor*, *Wolffia columbiana* and *W. arrhiza*. Fresh duckweeds were fed as a sole feed whereas dried duckweed meal was incorporated by partially replacing other conventional feed ingredients in pelleted diets. Feeding trials were conducted for varying periods, ranging from 60 to 155 days. Fish were fed *ad libitum* or at restricted level. In some studies,

TABLE 3.13
Performance of fish fed fresh duckweed

Duckweed/ Fish species	Rearing system	Rearing days	Control diet	Feeding rate (percent BW/day)		Fish size (g)	SGR (percent)	SGR as percent of control	FCR (DM basis[2])	Reference
				Fresh weight	Dry weight					
L. gibba/ tilapia hybrid[1]	Recirculatory tanks	89			>1.0	2.66	0.67		1.0	Gaigher, Porath and Granoth (1984)
L. gibba/ hybrid grass carp	Static water outdoor concrete tank	60	Catfish pellet (32 percent protein)	Ad libitum		1 015	0.21	56	6.69	Cassani, Caton and Hansen (1982)
L. perpusilla/ Nile tilapia	Static water outdoor concrete tank	70			2.5	26.3	0.97		2.2	Hassan and Edwards (1992)
					5.0	27.4	1.09		3.7	
L. perpusilla/ Nile tilapia	Static water outdoor concrete tank	70			3.0	43.7	1.34		1.6	Hassan and Edwards (1992)
					4.0	40.4	1.40		2.3	
L. minima/grass carp	Flow-through circular fibre glass tank	88		Ad libitum		2.7	3.74		1.72-1.97	Shireman, Colle and Rottmann (1977)
L. minima/grass carp	Flow-through circular fibre glass tank	68	Catfish chow (32 percent protein)	Ad libitum		2.8	3.88	212	1.6	Shireman, Colle and Rottmann. (1978)
						62.8	1.19	212	2.7	
S. polyrrhiza/ Nile tilapia	Static water outdoor concrete tank	70			2.5	25.6	0.59		3.1	Hassan and Edwards (1992)
					5.0	27.9	0.63		5.9	
W. columbiana/ hybrid grass carp	Static water outdoor concrete tank	60	Catfish pellet (32 percent protein)	Ad libitum		1 033	0.51	135	3.76	Cassani, Caton and Hansen (1982)
W. arrhiza/ grass carp	Cement cistern	133		Ad libitum		5.5	3.50			Naskar et al. (1986)
W. arrhiza/ silver carp	Cement cistern	133		Ad libitum		15.5	2.33			Naskar et al. (1986)
W. arrhiza/ common carp	Cement cistern	155		Ad libitum		15.0	1.90			Naskar et al. (1986)
W. arrhiza/ Java barb	Cement cistern	120		Ad libitum		9.5	2.49			Naskar et al. (1986)
W. arrhiza/ rohu	Cement cistern	155		Ad libitum		5.0	2.87			Naskar et al. (1986)
W. arrhiza/ mrigal	Cement cistern	155		Ad libitum		6.0	2.54			Naskar et al. (1986)

[1] *O. niloticus* X *O. aureus*
[2] Dry matter basis

TABLE 3.14
Performance of fish fed pelleted feeds containing dried duckweed

Duckweed/ Fish species	Rearing system	Rearing days	Control diet	Composition of test diet	Inclusion level (percent)	Fish size (g)	SGR (percent)	SGR as percent of control	FCR	Reference
L. minor/ Nile tilapia	Static water glass aquaria	70	Fishmeal: sesame oil cake: black gram bran: rice bran (27:25:19:25)	Duckweed meal incorporated by partial replacement of black gram and rice bran	13.5	2.55	2.13	98	2.1	Zaher et al. (1995)
L. minor/ Common carp	Static water outdoor cement cistern	140	Rice bran: groundnut cake (60:40) (17.5 percent protein)	Duckweed meal, groundnut cake, rice bran and ragi flour (40:20:20:20) (21.3 percent protein)	40	3.0	2.29	94	3.1	Devaraj, Krishna and Keshavappa (1981)

the performances of fish fed duckweed were compared with control diets, although in many of these studies no control diet was used for comparison.

Growth responses of different fish species fed various species of fresh duckweed were variable. However, the general trend was that the grass carp performed better than Nile tilapia and other species and the performances of duckweed as whole feed were better than control diet (Table 3.13). Similarly, duckweed meal incorporated in pelleted diets at 13.5 and 40 percent showed similar growth responses compared to the growth responses of fish fed control diets (Table 3.14). The SGRs obtained for grass carp fed fresh duckweed as whole feed varied between 1.2 and 3.9 while the SGR values for Nile tilapia were between 0.6 and 1.4. Fasakin, Balogum and Fasuru (1999) reported that duckweed meal (*Spirodela polyrrhiza*) can form up to 30 percent of the total diet of Nile tilapia without significant effect on performance, compared to a control without duckweed. However, inclusion levels above this level progressively decreased fish performance.

Duckweed are generally the preferred macrophytes for most of the herbivorous fish, although several authors reported that submerged macrophytes such as oxygen weed (*Hydrilla*) and water velvet (*Najas*) are more preferred than others. The preference of duckweed to other aquatic plants has been reported for grass carp and other fish species in several observations (Opuszynsky, 1972; Duthu and Kilgen, 1975; Rifai, 1979; Cassani, 1981; Cassani and Caton, 1983). Information on whether fish prefer any particular duckweed species over others is lacking.

Ad libitum feeding of fresh duckweed is mostly used for herbivorous fish. Limited numbers of investigations have been carried out to optimize the feeding or consumption rate of duckweed but most were carried out for grass carp and Nile tilapia. Nikolskij and Verigin (1966) reported grass carp consumed fresh duckweed equal to their body weight over a 24 hour period. Baur and Buck (1980) reported that grass carp consumed from 85 percent to 238 percent of their body weight/day (BW/day) on a mixed diet of *Lemna*, *Spirodela* and *Wolffia* spp. Shireman, Colle and Rottmann (1977) recorded consumption rates varying from 7.2-7.4 percent BW/day on a dry weight basis (DW) for grass carp while fresh duckweed (*L. minima*) was fed *ad libitum*. Since duckweed contains about 92 percent moisture, the dry weight feeding rates given above are equivalent to about 90-92 percent BW/day on a fresh weight basis. Shireman, Colle and Rottmann (1978) fed fresh *L. minima ad libitum* to grass carp and recorded daily mean consumption rates of 7.6 percent and 4.3 percent BW/day DW for 2.8 and 62.8 g sized fish respectively. Similar size-dependent feeding rates were reported by Hassan and Edwards (1992) for Nile tilapia. These authors studied the effect of feeding rate of *L. perpusilla* on the survival, growth and food conversion rate of Nile tilapia and recorded that the optimal daily feeding rates of *Lemna* were 5, 4 and 3 percent BW/day DW for fish of 25 to 44 g, 45 to 74 g and 75 to 100 g, respectively. Hassan and Edwards (1992) concluded that duckweed should be fed to tilapia according to its consumption rate, in order to avoid creating adverse water conditions, and that the feeding rate should be decreased as the fish grow larger.

A study by Effiong, Sanni and Sogbesan (2009) also indicated that the inclusion of duckweed meal in fish feeds could improve its binding potential and water stability.

3.5.2 Field studies and on-farm utilization

Several field studies and reports about the on-farm utilization of duckweed as feed for various fish species exist (e.g. Edwards, 1980, 1987; Edwards, Pacharaprakiti and Yomjinda, 1990; Skillicorn, Spira and Journey 1993; DWRP, 1998).

Edwards (1987) reported the on-farm utilization of duckweed in China and Taiwan Province of China. This author reported that the duckweeds *L. minor*, *S. polyrrhiza* and *W. arrhiza* are cultivated in small shallow ponds (similar to that illustrated in Figure 3.6) fertilized with manure (livestock or human) and fed to grass carp fry and fingerlings in nursery areas. Initially the fry are fed the smaller *Wolffia*, but when they reach 6 to 7 cm in length they are fed the larger *Lemna* and *Spirodela*. In Taiwan Province of China a mixture of *Lemna* and *Wolffia* is cultivated in shallow earthen ponds fed with faecally-polluted surface water for use as fish feed.

Skillicorn, Spira and Journey (1993) described the 'duckweed-fed carp polyculture system' developed in the PRISM experimental farm at Mirzapur, Bangladesh. The duckweed (*Lemna*, *Spirodela* and *Wolffia*) carp polyculture model has an 18-month cycle. Fingerlings were introduced in August and

FIGURE 3.6
Two boys collecting duckweed from a village pond (Jessore, Bangladesh)

September and harvesting began in March and continued for approximately one year. A second 18-month cycle began the following year and continued concurrently for six months. After the initial six months, the model allowed year-round harvesting. Bi-weekly harvesting was the preferred pattern, following a simple protocol to take the largest fish (75 to 100 percentile) and the smallest (0 to 25 percentile) in each species. The rationale was the assumption that the largest fish will exhibit a declining growth rate and that the small fish are simply poor performers.

The production rates achieved in this programme suggested that one hectare of duckweed production can support two hectares of carp polyculture. Empirical results suggested that a polyculture stocked at about 30 000 fish/ha may be fed as much duckweed as they will eat daily, regardless of the season. Fish were fed duckweed throughout the day. Freshly harvested duckweed was brought in baskets to the pond and distributed evenly among several 'feeding rings or squares' (Figure 3.7) consisting of 4 m² open-bottom enclosures. Feeding rings provide access by the fish to the duckweed and prevent it from dispersing over the pond surface. The feeding ring can be a floating enclosure anchored near the shore. Six feeding rings/ha were installed in the Mirzapur experimental site and appeared to provide sufficient access to food for all fish. Figures 3.8 and 3.9 show the transport and utilization of duckweed in fish culture.

In the Mirzapur experimental ponds, grass carp was the primary consumer of duckweed in the polyculture. However, both catla and common carp also competed aggressively for available duckweed feed and consumed it directly.

FIGURE 3.7
Duckweed cultivated in an undrainable pond (Mymensingh, Bangladesh)

FIGURE 3.8
A duckweed collector carrying a bag of duckweed in a rickshaw van. These professional duckweed collectors collect duckweed from various derelict ponds and sell them to fish farmers (Jessore, Bangladesh)

Top-feeders directly absorb about 50 percent of duckweed nutrients in their digestive systems. Their faeces contain the balance of the original duckweed nutrients and furnish relatively high quantity detritus to the bottom-feeders. A duckweed-fed fish pond thus appears to provide a complete, balanced diet for those carp that consume it directly, while the faeces of duckweed-feeding species, which are consumed directly by detritus feeders or indirectly through fertilization of plankton and other natural food organisms, provide adequate food for the remaining bottom and mid-feeding carp varieties. The fertilization of a duckweed-fed fish culture is therefore indirect and gradual, resulting from bacterial decomposition of fish faeces, dead algae, and other fermenting organic material.

FIGURE 3.9
A farmer is releasing a bag of duckweed to his carp pond. These duckweeds are generally purchased from a group of professional duckweed collectors (Jessore, Bangladesh)

Skillicorn, Spira and Journey (1993) reported that the first annual cycle of carp production produced slightly more than 10 tonnes/ha/year. However, these authors opined that a yield of between 10 to 15 tonnes/ha/year appears to be sustainable before biological constraints become limiting factors.

DWRP (1998) reported further follow-up of the duckweed-fed carp polyculture system developed by PRISM in Bangladesh. This report included the results of the demonstration farms as well as the results of the farmers' ponds. The duckweed-fed carp polyculture system practised by PRISM had two distinct differences from the model described by Skillicorn, Spira and Journey (1993). Apparently, PRISM included Nile tilapia with the traditional six-species carp culture system and provided other supplemental feed along with the duckweed. The farmers that adopted duckweed-based aquaculture produced an average of 3.6 tonnes/ha/year in comparison to the national average fish production of 2.1 tonnes/ha/year in Bangladesh at that time. PRISM itself achieved a production level of 11 tonnes/ha/year in 1993 and 16 tonnes/ha/year in 1996 in its demonstration farm.

The results of the duckweed-based carp polyculture of PRISM are presented in Tables 3.16, 3.17 and 3.18. Table 3.15 presents fish stocking and harvesting data in 1994. Tilapia were not stocked but multiplied on their own. Data for feed application, fish yield and food conversion ratio in farmers' ponds and in demonstration farms of PRISM are presented in Tables 3.16 and 3.17. The specific influence of duckweed feeding on yield and food conversion was obscured because other supplemental food was also added to the fish ponds. In neither demonstration ponds nor farmers' ponds could any example be found of pure duckweed feeding.

Table 3.17 presents data collected from demonstration farm production in Mirzapur over a period of three years. The results from fish ponds fed duckweed from organic wastewater plants have been kept separate from those fed on duckweed grown on chemical fertilizers. Whether there has been any difference in nutritive value between the duckweed from these different treatments could not be checked. What is interesting is to compare the difference in the ratio of duckweed to the other supplemental feed that was being applied. At first the information in Table 3.17 suggests that higher proportions of duckweed influence the conversion rate adversely. This conclusion may not be true, however, since the total amount of food applied in the ponds treated with waste-grown duckweed was clearly too high. Since the ponds did not show any increased production with a high rate of feeding, it must be assumed that they were at their carrying capacity most of the time and that all the extra food offered was apparently wasted. This seems to imply that the sustainable level of fish production from a duckweed-based polyculture lies around 10-15 tonnes/ha/year.

TABLE 3.15
The ratio of fish species stocked and harvested by PRISM in 1994

Species	Stocking rate (percent)	Harvest rate (percent)
Tilapia	0	38.8
Catla	20	6.7
Rohu	20	9.7
Mrigal	20	9.3
Silver carp	15	24.3
Grass carp	20	7.3
Common carp	5	3.3
Other	0	0.6

TABLE 3.16
Feed application, fish yield and food conversion ratio in farmers' ponds in two locations in Bangladesh during 1995-96

Location	Oil cake (kg/ha DM)	Wheat bran (kg/ha DM)	Duckweed (kg/ha DM)	Total (kg/ha DM)	Fish yield (kg/ha)	FCR
Tangail	2 742	1 441	1 526	5 833	3 290	2.1
Manikganj	2 556	1 854	1 465	5 874	5 007	1.5

Source: DWRP (1998)

TABLE 3.17
Feed application, fish yield and food conversion ratio in demonstration ponds at Mirzapur Experimental Site, Bangladesh during 1993-95

Nutrient source of duckweed	Season	Oilcake (kg/ha DM)	Wheat bran (kg/ha DM)	Duckweed (kg/ha DM)	Total (kg/ha DM)	percent duckweed used in feed	Fish yield (kg/ha)	FCR
Chemical	1993	8 504	1 065	6 662	16 231	41	13 430	1.2
	1994	11 722	507	5 902	18 131	33	15 080	1.2
	1995	12 107	122	5 810	18 039	32	11 520	1.6
Wastewater	1994	9 810	-	19 840	29 650	67	10 580	2.8
	1995	18 307	-	23 300	41 607	56	12 620	3.3

Source: DWRP (1998)

The food conversion values obtained when various duckweed species were fed to different fish species are presented in Table 3.18. Duckweeds were fed mostly in the fresh form and most of the values available are for grass carp and Nile tilapia. The values are variable, but the available data does not indicate if the variability was due to the fish species or to the duckweed species used. Generally, most FCRs are between 2.0 and 3.0, although an FCR of 1.0 was reported for hybrid tilapia and a very high FCR (6.7) for hybrid grass carp when both were fed *L. gibba*. This latter result was probably due to size of the fish (>1.0 kg) used in the feeding trial. Shireman, Colle and Rottmann (1978) reported an FCR of 1.6 for 2.8 g grass carp when fed fresh *L. minima* but the FCR was 2.7 for 63 g fish. Hassan and Edwards (1992) reported that food conversion was significantly affected by the feeding rate. For example, the FCR was 3.1 when *S. polyrrhiza* was fed to Nile tilapia at a feeding rate of 2.5 percent BW/day whereas it was 5.9 at a feeding rate of 5.0 percent BW/day. Similarly, FCR increased from 2.2 to 3.7 with an increase in feeding rate from 2.5 to 5.0 percent BW/day for Nile tilapia when fed *L. perpusilla*.

Generally, the FCR values reported for duckweed-based polyculture of carps in Bangladesh (Tables 3.16 and 3.17) were very good, being between 1.5 and 2.1 for farmer's ponds and 1.2 to 1.6 for demonstration ponds. However, it must be pointed out that duckweed was not used as the sole feed in these ponds, which were usually fertilized in addition to the use of oilcake, rice bran and wheat bran as supplemental feeds. Low FCRs for duckweed may be expected, since these plants have relatively low fibre and high protein contents (Table 3.12) and a good amino acid profile (Annex 1 Table 3). Although it is difficult to generalize from the available data, an FCR value of 2.5 may be a reasonable expectation for grass carp and Nile tilapia based.

TABLE 3.18
Food conversion ratios of duckweed to fish

Duckweed	Fish species	Fish size (g)	Food conversion ratio (FCR)		Reference
			DM[2]	FW[2]	
L. gibba, fresh	Tilapia hybrid	2.7	1.0		Gaigher, Porath and Granoth (1984)
L. gibba, fresh	Hybrid grass carp	1 015	6.7		Cassani, Caton and Hansen (1982)
L. perpusilla, fresh	Nile tilapia	8-10	3.7	60.6	Edwards, Pacharaprakiti and Yomjinda (1990)
L. perpusilla, fresh	Nile tilapia	26-27	2.2-3.7		Hassan and Edwards (1992)
L. perpusilla, fresh	Nile tilapia	40-44	1.6-1.9		Hassan and Edwards (1992)
L. minima, fresh	Grass carp	2.7	1.7-2.0		Shireman, Colle and Rottmann (1977)
L. minima, fresh	Grass carp	2.8	1.6		Shireman, Colle and Rottmann (1978)
L. minima, fresh	Grass carp	63	2.7		Shireman, Colle and Rottmann. (1978)
L. minor, fresh	Nile tilapia	n.s.		33	Rifai (1979)
L. minor, dried	Nile tilapia	2.5	2.1		Zaher *et al.* (1995)
L. minor, dried	Common carp	3.0	3.1		Devaraj, Krishna and Keshavappa (1981)
Lemna sp., fresh	Grass carp	n.s.		37	Hepher and Pruginin (1979)
S. polyrrhiza, fresh	Nile tilapia	26-28	3.1-5.9		Hassan and Edwards (1992)
S. polyrrhiza, dried, 30 percent inclusion	All-male tilapia	13.9	2.0		Fasakin, Balogun and Fasuru (1999)
W. arrhiza, fresh	Six carp species[1]	5.0-15.5	5.6	78.8	Naskar *et al.* (1986)
W. columbiana, fresh	Hybrid grass carp	1 033	3.8		Cassani, Caton and Hansen (1982)
Mixture of *Lemna*, *Spirodela* and *Wolffia*	Grass carp	n.s.	1.6-4.1		Baur and Buck (1980)

[1] Polyculture of six carp species (grass carp, silver carp, common carp, Java barb, rohu and mrigal)
[2] FW = fresh weight basis; DM = dry matter basis

Digestibility coefficients of *Lemna*, *Spirodela* and *Wolffia* fed to grass carp, tilapia and rohu are presented in Table 3.19. Considering the importance of duckweed as fish feed, it is surprising to note that only a few studies have been carried out to investigate its digestibility for fish. Van Dyke and Sutton (1977) were probably the first to investigate the digestibility of duckweed (mixture of *L. minor* and *L. gibba*) in detail for grass carp. These authors estimated the true dry matter digestibility of duckweed to be 65 percent, while the apparent digestibility was 53 percent for dry matter, 80 percent for crude protein, 58 percent for organic matter, 26 percent for ash and 61 percent for gross energy. The dry matter digestibility of *L. gibba*, *S. polyrrhiza* and *W. arrhiza* for grass carp found by other authors (Table 3.19) varied between 67-82 percent, while the dry matter digestibility of *L. gibba* for hybrid tilapia was reported to be 65 percent. Grass carp passes its food rapidly through a short, unspecialized gut and the fish probably does not produce cellulase (Van Dyke and Sutton, 1977); it is therefore unrealistic to expect that more than 50-60 percent of the feed consumed would actually be digested.

TABLE 3.19
Digestibility of duckweed for selected fish species

Duckweed	Fish species	Fish size (g)	Digestibility (percent)[1]						Reference
			DM	CP	EE	NFE	CF	GE	
L. gibba	Hybrid tilapia[2]	2.7	65	86					Gaigher, Porath and Granoth (1984)
L. gibba	Grass carp		82						Lin and Chen (1983, cited by Wee, 1991)
L. gibba and *L. minor* (1:1)	Grass carp	320	53	80				61	Van Dyke and Sutton (1977)
S. polyrrhiza	Grass carp		75						Lin and Chen (1983, cited by Wee, 1991)
W. arrhiza	Grass carp		67						Lin and Chen (1983, cited by Wee, 1991)
W. arrhiza	Rohu	3.6		91.5	93.5	81.2		84.4	Ray and Das (1994)

[1] DM = dry matter; CP = crude protein; EE = ether extract; CF = crude fibre; NFE = nitrogen free extract; GE = gross energy
[2] *O. niloticus* X *O. aureus*

4. Floating aquatic macrophytes – Water hyacinths

Mature plants of water hyacinths (*Eichhornia crassipes*) consist of long, pendant roots, rhizomes, stolons, leaves, inflorescences and fruit clusters. The plants may be up to 1 m high, although 40 cm is the more usual height. The inflorescence bears 6-10 lily-like flowers, each 4-7 cm in diameter. The stems and leaves contain air-filled tissue, which gives the plant considerable buoyancy. Vegatative reproduction takes place at a rapid rate under preferential conditions (Herfjord, Osthagen and Saelthun, 1994).

Water hyacinths are considered as nuisance species because they multiply rapidly and clog lakes, rivers and ponds. The thick mats (Figure 4.1) formed

FIGURE 4.1
Part of River Yamuna covered with lush green water hyacinth, Delhi, India

Courtesy of Rina Chakrabarti

under favourable conditions often obstruct fishing, shipping and irrigation and are hard to eradicate. Great efforts are being made to contain water hyacinths but, on the other hand, attempts are being made to find practical uses for the large biomass that is available. It offers the potential for use as fodder for domestic animals, as fish feed, for the production of biogas and for the removal of heavy metals and phenols from polluted waters. For example, studies have shown that about 1 million L/day of domestic sewage could be treated over an area of 1 ha through water hyacinths, reducing the BOD and COD by 89 and 71 percent, respectively (Reddy *et al.*, 2005).

4.1 CLASSIFICATION
There are seven species of water hyacinth, the best known being the common water hyacinth, *Eichhornia crassipes*, which is a perennial free-floating aquatic plant belonging to the family Pontederiaceae (Figure 4.2).

4.2 CHARACTERISTICS
4.2.1 Importance
Water hyacinths are found in most of the tropical and subtropical countries of the world. According to Mitchell

FIGURE 4.2
Common water hyacinth (*Eichhornia crassipes*)

Source: USDA

(1976), the water hyacinth is indigenous to South America, particularly to the Amazonian basin. It started its worldwide journey as an ornamental plant when first introduced into the USA in 1884 (Penfound and East, 1948 cited by Edwards, 1980). It reached Australia in 1895, India in 1902, Malaysia in 1910, Zimbabwe in 1937 and the Republic of the Congo in 1952.

4.2.2 Environmental requirements

According to Wilson *et al.* (2001) there are five main factors limiting the growth rate and carrying capacity of water hyacinth: salinity, temperature, nutrients, disturbance and natural enemies.

Water hyacinths seem unable to survive salinities above 2 ppt. Olivares and Colonnello (2000) reported that water hyacinth survives salinities of 1.3-1.9 ppt in the Orinoco (South America) and Kola (1988) reported that the plant grew well at salinities below 1 ppt.

Low temperatures stop the plant establishing in temperate areas and prevent it from reaching high levels in the sub-tropics. Knipling, West and Haller (1970) proposed a parabolic relationship between temperature and growth rate, with growth tailing off quickly after the optimum of 30 °C. Imaoka and Teranishi (1988) proposed that the intrinsic growth rate, r, increases exponentially with ambient temperatures in the range 14-29 °C, growth ceasing below 13 °C. Frost is a major cause of leaf mortality in temperate regions. Applying mathematical modelling, using existing data, Wilson, Holst and Rees (2005) examined the role of two important environmental factors, temperature and nutrient level, on the growth of water hyacinths. Their model predicted a linear reduction in specific growth rate with density. These authors set the minimum (Q_{min}), optimum (Q_{opt}) and maximum (Q_{max}) temperatures for water hyacinth as 8, 30 and 40 °C, respectively. The growth of water hyacinths is affected by low air humidity, ranging from 15-40 percent relative humidity (Freidel and Bashir, 1979).

The levels of available nitrogen and phosphorous are the most important factors limiting growth (Wilson *et al.* 2001). The half-saturation co-efficients for water hyacinths grown under constant conditions have been found to be from 0.05-1 mg/l for total nitrogen and from 0.02-0.1 mg/l for phosphates. Growth quickly tails off below the lower limits. Wilson, Holst and Rees (2005) suggested that nitrogen is limiting if total nitrogen concentration is less than seven times that of the phosphorus concentration. Water hyacinths show logistic growth. The model assumed that plants grow in the absence of interspecific competition. In fact, the plant soften grow in areas previously free of aquatic vegetation.

Flooding can break up large mats of water hyacinth and leave plants stranded on land (Wilson *et al.* 2001). Wave action may limit growth by directly damaging plants and by forcing the weed to maintain aerenchymatous tissue.

4.3 PRODUCTION

Water hyacinth grows in all types of freshwater, lentic and lotic. Westlake (1963) predicted that water hyacinths might be exceptionally productive plants since they are warm water species with submerged roots and aerial leaves like emergent macrophytes. Production statistics of this macrophyte in various aquatic environments are available (Table 4.1). The productivity varies widely and is dependent on the environment under which it grows. Wolverton and McDonald (1976) reported a yield of water hyacinth of up to 657 tonnes/ha/year DM in ponds fertilized with sewage nutrients, while Coche (1983) reported an even higher yield of 750 tonnes/ha/year in irrigation canals in China. However, many of these reported yields are extrapolated. It may therefore not be possible to obtain the higher calculated productivities on a large scale, since it would be difficult to maintain the most rapid growth rates obtained on a small experimental scale throughout the year (Edwards, 1980). The latter author, however, opined that an annual production of 200 tonnes/ha/year might be attainable in eutrophic waters in the tropics.

TABLE 4.1
Productivity of water hyacinths under different aquatic environments

Aquatic environment	Yield (tonnes/ha/year)
Fertile ponds	15-200
Artificially fertilized ponds	75.6-191.1
Fertilized pond	70.8
Fertilized pond with sewage effluent	212-657
Fertilized pond with sewage effluent	219
Irrigation canals in China	400-750
Nutrient non-limiting water of Florida, USA	106
Man-made lakes of central Java	255

Source: Edwards (1980); Little and Muir (1987)

China is probably the only country where water hyacinth has been reported to be cultivated with two other aquatic macrophytes, namely water lettuce (*Pistia stratiotes*) and alligator weed *Alternathera philoxeroides* (Edwards, 1987). These plants are usually cultivated in rivulets, small bays or swamps, and are usually fed to pigs.

4.4 CHEMICAL COMPOSITION

A summary of the chemical composition of water hyacinths (fresh, dried and composted) from different geographic regions of the world is presented in Table 4.2. Like most other aquatic macrophytes, water hyacinths have very high moisture content; the dry matter generally varies between 5-9 percent. Table 4.2 indicates that there is little variation in proximate composition in relation to geographic location. Variation, however, does exist between the proximate composition of whole plants and leaves. The crude protein content of the whole plant is about 12-20 percent DM, although a level as low as 9 percent was reported in studies. Gohl (1981) reported that the crude protein of fresh green part of water hyacinths from India and the Philippines was 12.8-13.1 percent DM. The crude protein content of leaf meal appears to be higher than the whole plant and varies between 20-23 percent.

Like most other aquatic macrophytes, the crude lipid content of water hyacinths is usually low and varies between 2-4 percent on dry matter basis regardless of whole plant or leaves. The ash content of whole plants varies between 15-34 percent while it is between 10-18 percent for leaves. Crude fibre content is usually high in water hyacinths and ranges between 17-32 percent, irrespective of whole plant or leaves. Some information on the amino acid content of various aquatic macrophytes is contained in Annex 1.

Gunnarsson and Petersen (2007), in a review that covered water hyacinths collected from various sources, also reported levels of some other components: hemicellulose 22-43.4 percent; cellulose 17.8-31 percent; lignin 7-26.36 percent; and magnesium 0.17 percent. Matai and Bagchi (1980) provided some additional component levels for fresh water hyacinths, namely that the ash contained 28.7 percent K_2O, 1.8 percent Na_2O and 21 percent Cl.

4.5 USE AS AQUAFEED

Because of their relatively high protein content and abundance in tropical and sub-tropical countries, a significant number of research studies have been carried out to find the potential for the utilization of water hyacinths as a fertilizer, for example by Sipauba-Tavares and Braga (2007) for the rearing of tambaqui (*Colossoma acropomum*), and as a fish feed in pond aquaculture. Available literature indicates that water hyacinths are fed to fish either in fresh form, or as a dried meal in pelleted diets, or composted as feed and fertilizer. Apart from these three forms, attempts are

TABLE 4.2
Proximate composition and mineral content of water hyacinths

Form	DM (%)	Proximate composition[1] (% DM)					Minerals (% DM)		Reference
		CP	EE	Ash	CF	NFE	Ca	P	
Fresh whole plant, Vietnam	5.8	13.4	2.5	34.1	31.8	18.2	n.s.	n.s.	Tuan et al. (1994)
Dried whole plant, Thailand	89.6	15.8	3.9	30.3	22.8	27.2[3]	n.s.	n.s.	Edwards, Kamal and Wee (1985)
Whole plant, sun-dried, Thailand	90.2	8.9	1.0	17.3	23.0	49.8	n.s.	n.s.	Klinavee, Tansakul and Promkuntong (1990)
Fresh leaf, Thailand	7.3	22.7	3.9	18.8	17.5	37.1[3]	1.79	0.83	Somsueb (1995)
Leaf meal, Bangladesh[2]	89.3	29.7	3.6	10.2	17.2	39.3	1.15	0.64	Hasan (1990)
Leaf meal, India	91.8	22.8	0.8	13.1	19.4	43.9	n.s.	n.s.	Murthy and Devaraj (1990)
Leaf meal, India	90.8	19.7	2.8	10.2	20.0	47.3[3]	n.s.	n.s.	Nandeesha et al. (1991)
Whole plant, composted, Thailand	89.5	15.8	1.4	49.8	10.4	22.6[3]	n.s.	n.s.	Edwards, Kamal and Wee (1985)
Leaf protein concentrate, USA	n.s.	32.1	7.7	7.3	7.9	45.0	n.s.	n.s.	Liang and Lovell (1971)
Fresh whole plants	4.5	9.4	n.s.	24.2	n.s.	...	2.21	0.74	Matai and Bagchi (1980)
Fresh whole plants, various locations	6.2–9.4	11.9–20.0	3.5	15.0–25.7	18.9	...	0.58	0.53	Gunnarsson and Petersen (2007)
Mean from various sources	8.1	15.5	1.9	17.8	25.3	39.5[3]	2.06	0.6	Hertrampf and Piedad-Pascual (2000)

[1] CP = crude protein; EE = ether extract; CF = crude fibre; NFE = nitrogen free extract; Ca = calcium; P = phosphorus
[2] Leaves collected from a particular type of water hyacinth with a long and thin stem
[3] Adjusted or calculated; not as cited in original publication

also made to feed water hyacinths to fish by processing them with other techniques. Many of these studies were conducted under laboratory conditions and reports of on farm utilization as fish feed are rather limited. Information on these topics has been grouped into several sections: four dealing with the various forms of water hyacinth (fresh, dried, composted and fermented, and other processing techniques), followed by comments on food conversion efficiency and digestibility.

4.5.1 Fresh form

Many researchers have investigated the use of water hyacinth in its fresh form. The high moisture content is a major constraint in its use as fish feed, which has proved to be unsuccessful in many cases. Hyacinth leaves are generally cut into small pieces and fed to grass carp or other macrophytophagous fish. Generally, grass carp feed on this plant only when no other macrophytes or feeds are available.

Riechert and Trede (1977) reported the results of a preliminary indoor laboratory trial carried out in Germany on the feeding of water hyacinths to grass carp. Eleven month old fish weighing 38 to 104 g were fed for 50 days exclusively on water hyacinths. Roots and leaves were accepted readily by the grass carp but the swollen petioles reluctantly. The fish grew well, producing 6.5 g live weight from 10 g DM hyacinth (FCR = 1.54). These authors also noted that grass carp above 80-100 g were better able to utilize hyacinth leaves compared to smaller fish and postulated that only 50-60 percent of the feed consumed was actually digested.

Tuan *et al.* (1994) used both fresh and fermented water hyacinth as supplementary feed in nursery ponds in Vietnam for fingerlings (1-6 g) of Nile tilapia, common carp, grass carp and Java barb. Fresh whole water hyacinth was chopped and mixed with rice bran at a ratio of 2:1 or 1:1 and fed to fish. A water hyacinth-rice bran mix was also fermented and fed. The growth of fish obtained by feeding the hyacinth-rice bran mixture was comparable to the growth obtained from rice bran alone. Rice bran is normally applied to nursing ponds in Vietnam. In terms of weight gain and specific growth rate, water hyacinths mixed with rice bran at a ratio of 2:1, either raw or fermented, could be used to replace rice bran in nursery ponds. Amongst the four species used, Nile tilapia performed better than the other species, exhibiting a specific growth rate of 4.3-4.8 percent/day. The specific growth rates of grass carp, Java barb and common carp were 4.06-4.19 percent, 2.84-3.00 percent and 2.49-2.66 percent per day, respectively.

As noted above, the use of fresh water hyacinth as fish feed has achieved limited success, principally because of its high moisture content. There are several other limitations to its use for this purpose. For example, the fresh plant contains prickly crystals, which make it unpalatable (Gohl, 1981). This was thought to be probably due to the presence of raphids and oxalates in water hyacinths (Dey and Sarmah, 1982). Microscopic examination of water hyacinths reveals the presence of sharp needles formed by calcium oxalate, which may be harmful for fish (Bolenz, Omran and Gierschner, 1990).

Fresh whole water hyacinth has been applied to ponds as feed and fertilizer in China, but the fish were reluctant to accept it and it took a long time to decompose, eventually resulting in inefficient utilization (Anonymous, 1980). Several processing techniques have therefore been employed to increase its nutritive value and to decrease the high moisture content. These include its use in dried and composted forms, and the incorporation of leaf meal in pelleted feeds. Another practice prevalent in China is the application of paste or mashed water hyacinth, which releases the mesophyll cells in water for consumption by carps. The processing methods employed so far and the results achieved with various fish species are summarized in subsequent sections.

4.5.2 Dried meal form

One of the most commonly used methods for processing of water hyacinth is drying. In tropical and sub-tropical countries, water hyacinths are often sun-dried, as other drying methods can be expensive. Two days of good sun drying would be sufficient to reduce the moisture content to about 10-12 percent. A number of growth studies have been conducted under laboratory conditions using dried water hyacinth in pelleted feeds for carps, tilapia and catfish. In most cases the dried water hyacinth was ground into a meal and fed to fish, partially or completely replacing fishmeal or other conventional protein sources.

A summary of the results of the selected growth studies carried out on the use of dried water hyacinth meal in pelleted feeds for different fish species is presented in Table 4.3. Whole water hyacinth or its leaf meal was evaluated as a major ingredient in pelleted diets for Nile tilapia (*Oreochromis niloticus*), Java tilapia (*O. mossambicus*), grass carp (*Ctenopharyngodon idella*), common carp (*Cyprinus carpio*), the Indian major carp rohu (*Labeo rohita*), stinging catfish (*Heteropneustes fossilis*), Java barb (*Barbonymus gonionotus*), sepat rawa (*Trichogaster* sp.), matrincha (*Brycon* sp.) and African catfish (*Clarias gariepinus*). The dietary incorporation level of water hyacinth meal used varied widely, ranging from as low as 2.5 percent to as high as 100 percent. In most of these studies, the performance of fish fed diets containing various inclusion of water hyacinth was compared with the use of control diets. Various types of control diets were used, including commercial pellets, fishmeal-based pellets, the traditionally used rice bran-oil cake mixtures, and a mixture of fishmeal and cereal by-products.

Growth responses of different fish species fed test diets containing different inclusions of water hyacinth meal have been highly variable. For example, significant reduction in growth responses were reported by Hasan, Moniruzzaman and Omar Farooque (1990) for rohu fry and by Hasan and Roy (1994) for rohu fingerling when 27-30 percent water hyacinth leaf meal was included to replace the fishmeal protein of the control diet. Similarly, Klinavee, Tansakul and Promkuntung (1990) recorded significant reduction in growth responses of Nile tilapia when fed a test diet containing 40 percent water hyacinth meal. However, Murthy and Devaraj (1990), using a 50 percent dietary inclusion level in diets for grass carp and common carp, Dey and Sarmah (1982) using 100 percent inclusion for Java tilapia, and Saint-Paul, Werder and Teixeira (1981), using 18.5 percent inclusion for matrincha (*Brycon* sp.), respectively recorded either similar or higher growth responses compared to control diets. However, in some of these studies, the control diet consisted only of a rice bran-oil cake mixture, which may itself have not generated good growth. Edwards, Kamal and Wee (1985) tested the growth response of Nile tilapia to 75 and 100 percent displacement of a 32.5 percent protein commercial tilapia pellet by water hyacinth meal. The test diets resulted in only a 10-15 percent reduction in SGR. This is an interesting performance for water hyacinth meal. However, these authors concluded that although the experimental fish obtained their nutrition directly from the diets, they must also have obtained some indirect nutrition from the plankton in the static water experimental system used. This assumption of indirect nutritional benefit from phytoplankton may also have been true in the experimental studies conducted by Dey and Sarmah (1982) and Murthy and Devaraj (1990).

Hertrampf and Piedad-Pascal (2000) suggested inclusion rates for water hyacinth in farm-mixed feeds for the farming of herbivorous or omnivorous freshwater fish in simple farming systems where it is available at low cost. These authors recommended that suitable inclusion levels were either 25-50 percent as a supplementation of basic feed (e.g. rice bran, broken rice, chicken manure) or 5-10 percent as a replacement protein source in formulated feeds (fish meal, vegetable oil meals/cake).

TABLE 4.3
Performance of different fish species to pelleted feeds containing dried water hyacinth meal

Fish species	Rearing system	Rearing days	Control diet	Composition of test diet	Dietary WH inclusion level (%)	Fish size (g)	SGR (%)	SGR as % of control	FCR	Reference
Nile tilapia	Static water outdoor concrete tank	84	Commercial pellet (32% protein)	WH meal incorporated at two levels in control diet by simple displacement	75 100	14.44 13.50	1.80 1.70	90.5 85.4	2.41 2.61	Edwards, Kamal and Wee (1985)
Nile tilapia	Clear water fibre glass tank	77	Chicken pellet (16.8 protein)	WH leaf meal, fishmeal and rice bran (40:18.5:34.3)	40	7.70	1.01	64.0	4.3	Klinavee, Tansakul and Promkuntong (1990)
Java tilapia	Indoor static glass aquaria	28	Rice bran: mustard oil cake (50:50)	Dried WH petiole meal	100	15–22	–[2]	n.s.	n.s.	Dey and Sarmah (1982)
Java tilapia	Cages in lake	70	No control diet	WH meal, fishmeal, corn meal and rice bran[1]	2.5 10.0	10–20 10–20	1.34 1.30	n.s n.s	n.s n.s	Hutabarat, Syarani and Smith (1986)
Grass carp	Static water outdoor cement cistern	112	Rice bran: groundnut cake (50:50)	WH leaf meal, groundnut cake, rice bran and fishmeal (50:21:15:10)	50	6.53	2.58	108.9	2.79	Murthy and Devaraj (1990)
Common carp	Static water outdoor cement cistern	112	Rice bran: groundnut cake (50:50)	WH leaf meal, groundnut cake, rice bran and fishmeal (50:21:15:10)	50	3.13	3.21	107.7	3.68	Murthy and Devaraj (1990)
Common carp	Cages in lake	70	No control diet	WH meal, fishmeal, corn meal and rice bran[1]	2.5 10.0	10–20 10–20	1.61 1.51	n.s n.s	n.s n.s	Hutabarat, Syarani and Smith (1986)
Rohu	Indoor static glass aquaria	42	Fishmeal based diet (40% protein)	20 and 40% of total dietary protein from fishmeal replaced by WH leaf meal	27 54	0.20 0.21	2.50 2.21	79.9 70.6	2.31 2.36	Hasan, Moniruzzaman and Omar Farooque (1990)
Rohu	Indoor static glass aquaria	77	Fishmeal based diet (30% protein)	25 and 50% of total dietary protein from fishmeal replaced by WH leaf meal	30 60	3.50 3.50	0.90 0.79	63.7 65.3	2.98 3.31	Hasan and Roy (1994)
Stinging catfish	Indoor static glass aquaria	14	Minced meat (87% of diet)	Minced meat and WH leaf meal (37:50)	50	10.5	–[3]	n.s	2.2	Niamat and Jafri (1984)
Java barb	Cages in lake	70	No control diet	WH meal, fishmeal, corn meal and rice bran[1]	2.5 10.0	10–20 10–20	0.96 1.44	n.s n.s	n.s n.s	Hutabarat, Syarani and Smith (1986)
Sepat rawa	Cages in lake	70	No control diet	WH meal, fishmeal, corn meal and rice bran[1]	2.5 10.0	10–20 10–20	1.36 0.93	n.s n.s	n.s n.s	Hutabarat, Syarani and Smith (1986)
Matrincha	Suspended cloth pond connected to a water treatment plant	90	Fishmeal, wheat meal and corn meal (27:35:28)	Two levels of WH meal incorporated by replacing similar amount of corn meal from the control diet.	9.5 18.5	1.5 1.5	1.0 0.9	125.0 112.5	1.7 1.8	Saint-Paul, Werder and Teixeira (1981)
African catfish	Plastic tanks	70	Fishmeal based diet (35% protein)	WH meal replaced fishmeal at four levels	10 20 30 40	1.26 1.26 1.26 1.26	0.64 0.67 0.63 0.54	76.2 79.8 75.0 64.2	3.5 3.3 3.4 3.6	Konyeme, Sogbesan and Ugwumba (2006)

[1] One test diet contained 2.5% WH meal with fishmeal, corn meal and rice bran in the ratio of 35.0:10.0:52.5. The other test diet contained 10% WH meal with 7.5% reduction of rice bran
[2] SGR not reported. Consumption of control diet was 72%, while it was 70% for test diet. The performance in terms of weight gain was similar for fish fed control and test diet
[3] SGR not reported. Fish fed water hyacinth diet registered around 20% gain in live weight in contrast to the group fed control diet where weight gain was about 8%

4.5.3 Composted and fermented forms

Composting or fermentation are techniques often used to reduce water hyacinth into forms utilizable for feeding livestock.

Composting is one of the most widely used processing techniques to prepare water hyacinth for use as a fertilizer or fish feed (Figure 4.3). A large quantity of inorganic nitrogen and phosphorus accumulates in the roots of water hyacinth, which makes it suitable as a compost or inorganic fertilizer. However, a major problem with the use of water hyacinth meal in fish diets is its relatively high crude fibre content. Fish do not appear to produce cellulase directly (Buddington, 1980) and their ability to maintain a symbiotic gut flora capable of hydrolyzing cellulose is limited. Fish often poorly accept water hyacinth leaf meal in pelleted diets. This has been identified as one of the major contributory factors for the reduced growth responses of major carp (*L. rohita*) fry by Hasan, Moniruzzaman and Omar Farooque (1990). Composting has been reported to increase the nutritive value and acceptability of water hyacinth. Edwards. Kamal and Wee (1985) made a comparison of the proximate composition of composted water hyacinth and dried water hyacinth meals and observed that while the crude protein levels were similar, the crude fibre and crude fats levels were approximately halved and the ash content approximately doubled by the composting process.

FIGURE 4.3
Two farmers carrying dry water hyacinth to the pond side for preparation of compost pit (Mymensingh, Bangladesh)

Preparation and use of composted water hyacinth

The most commonly used method for compost preparation is the Chinese method of surface continuous aerobic composting. Edwards. Kamal and Wee (1985) described the method as follows. Whole water hyacinth plants are cut into 2-3 cm pieces by a rotary chopper and sun-dried to an ambient equilibrium moisture content of about 20 percent on a platform elevated above the ground to facilitate drying. Compost is made by mixing dried and freshly chopped water hyacinth to give an initial pile moisture content of 65-70 percent; the mixture is made into a pile 2.5 m (length) x 2 m (width) x 1.3 m (height) and perforated bamboo poles are inserted for aeration. The mixture is turned occasionally to facilitate decomposition. The composting process is completed within 50 days.

Urea is often added at 2 percent to speed up the decomposing process. In this process it is suggested that the compost should be prepared by mixing water hyacinth, cow dung, urea and lime; water hyacinth and cow dung constituting the bulk of the ingredients while urea and lime are added at 2-5 percent of the total. The ingredients are kept in an earthen pit and arranged in layers with the top covered by polythene, paper or banana leaves (Figures. 4.4 and 4.5). Perforated bamboo poles are inserted for aerobic decomposition. However, compost preparation has been reported to be labour intensive and farmers are often reluctant to prepare compost for use as fertilizer. A simple compost preparation technique for use in fish ponds has been developed by the Mymensingh Aquaculture Extension Project (MAEP, Bangladesh) by using water hyacinth, cow dung, urea and lime (M.A. Mirza, MAEP *pers com.* 2004). Freshly procured whole water hyacinths are chopped into small pieces and dried for 1-2 days in sunlight. Sun-dried water hyacinth containing about 15-20 percent moisture is mixed with cow dung, lime and urea in the ratio of 88:10:1:1 (water hyacinth: cow dung: lime: urea). The ingredients are not kept in layers as traditionally used but are thoroughly mixed. The mixture is kept for decomposition in a pit near the pond side. The mixture is re-mixed every seven days to facilitate decomposition. The minimum area of the

FIGURE 4.4
A compost pit prepared with water hyacinth and cow dung (Mymensingh, Bangladesh)

compost pit is 1 ft^2, with a recommended depth of 4 ft to hold 70 kg of compost. The composting process is normally completed within two months. The recommended rate of compost application as suggested by MAEP is 18 000 kg/ha/year.

Compost is traditionally used as fertilizer in fish ponds in many Asian countries. Reports on its use as a fish feed are rather limited, however. Composted water hyacinth was evaluated as fish feed in pelleted diet for Nile tilapia by Edwards, Kamal and Wee (1985). These authors prepared four test diets by incorporating 25, 50, 75 and 100 percent of composted water hyacinth meal in a control diet that consisted of a conventional pelleted tilapia feed (32 percent protein). Good growth and feed utilization efficiencies were obtained with diets containing up to 75 percent composted water hyacinth, with no significant reduction in fish performance compared to the control diet. The specific growth rates varied between 1.96 and 2.15 for test diets while the SGR for control diet was 1.99. The FCR was between 2.18 and 2.57 for the diets with compost and 2.63 for the control.

Similarly Hutabarat, Syarani and Smith (1986) reported good growth by using composted water hyacinth in a pelleted feed for Java tilapia, Java barb and common carp in cage culture. However, these authors used only 10 percent as their maximum inclusion level. Edwards (1987) reported that good results were obtained in China by composting water hyacinth with silkworm faeces (or animal manure) and quicklime, or by composting the chopped water hyacinth with a small amount of salt or saccharified yeast.

The *in situ* decomposition of water hyacinth and its efficacy was studied by Mishra, Sahu and Pani (1988)

FIGURE 4.5
View of a compost pit in a corner of a pond (Mymensingh, Bangladesh)

in rearing ponds for Indian major carp (rohu, catla and mrigal) fingerlings. Fresh water hyacinth was applied at 300 kg/month/0.2 ha pond (1 500 kg/ha/month). The water hyacinth was killed *in situ* by using an aqueous solution of 2,4-D sodium salt and was allowed to decompose and disintegrate in the pond. The ponds were stocked at a rate of 3 000 fingerlings per ha (600/pond) and reared for twelve months. The addition of water hyacinth increased the fish production by about 52 percent as compared to the control pond where no additional input was provided. A net increase of 64.7 kg of fish was obtained by using 3 600 kg of fresh water hyacinth. The conversion ratio worked out to be ~55.7 for fresh hyacinth, while the FCR was about 3.3 on a dry matter basis, considering that fresh hyacinth contains about 6 percent DM.

Rohu (*Labeo rohita*) larvae were stocked at 1 million/ha by Sahu, Sahoo and Giri (2002) under three culture conditions: the application of water hyacinth compost (8 000 kg/ha), inorganic fertilizer (60 kg/ha), or no manure (control). While the total nitrogen and phosphate levels of the control treatment were 0.02 and 0.04 g/L, those in the compost treatment were 0.17 and 0.08 mg/l after 15 days of fertilization. In the inorganic fertilizer treatment the nitrogen level was elevated to 0.12 mg/l after 15 days but the phosphate level remained at 0.04 mg/l throughout the study period. The plankton volumes were 1.8, 1.2 and 0.4 ml/45 L in the compost, fertilizer and control treatments, respectively at the time of stocking of larvae. Significantly ($P < 0.01$) higher survival and growth were found in the compost treatment compared to other treatments.

Preparation and use of fermented water hyacinth

Fermentation is an age-old practice in food processing. In many cases fermentation has been reported to improve the nutritive value of cereal grains and oilseeds by increasing their protein efficiency ratio, digestibility and the availability of free amino and fatty acid contents. During fermentation, nutrient losses may occur as a result of leaching, destruction by light, heat or oxygen, or microbial utilization (Jones, 1975). Nevertheless, the loss of nutrients during this process is generally small and there may even be an increase in the nutrient level through microbial synthesis.

Edwards (1987) reported that water hyacinths were processed in China either mechanically (soaking, mixing, cutting, or grinding) or biologically for feeding to grass carp and common carp. The biological processing involved green storage and fermentation in ditches, tubs, or barrels under anaerobic conditions at 65-75 percent moisture after cutting into 6 cm strips and sealing by a 15 cm layer of dry grass topped by a 15 cm layer of moist soil; if the material was too moist it could be sun-dried or mixed with dry hay before sealing.

A simple fermentative treatment with cow dung and urea was evolved to process and utilize water hyacinth, as a feed and manure for carp culture by Olah, Ayyappan and Purushothaman (1990). Water hyacinth leaves were chopped into 5 cm pieces and mixed with 10 percent cow dung and 2 percent urea. The mixture was then kept in an airtight polystyrene bag and incubated at room temperature (27-32 °C). These authors observed that a period of 2-3 weeks was optimal for cellulose degradation and to improve the nutritive value of water hyacinth. The crude protein content of the substrate increased from 13.1 to 18.1 percent of the dry weight during 18 days of treatment.

Olah, Ayyappan and Purushothaman (1990) fed fermented water hyacinth to catla, rohu, mrigal, silver carp and common carp in trials conducted in plastic pools for four (Trial 1) and eight week (Trial 2) periods. The stocking density was 19 and 5/m^2 for Trials 1 and 2, respectively, with daily feeding rates of 50 g/m^2. Silver carp and mrigal showed the best growth rates, followed by rohu. Food conversion ratios of 2.02 and 3.72 were obtained for Trials 1 and 2, respectively. Fermentation of water hyacinth may thus be a simple and efficient treatment for utilizing water hyacinths as a feed or manure in fish culture without the energy-consuming process of pelletization.

Xianghua (1988) reported on the use of fermented water hyacinth as feed for grass carp. The plant was harvested, chopped, blended with a small amount of corn flour and fermented overnight. Good results were obtained in rearing grass carp beyond age II+.

El Sayed (2003) reported that Nile tilapia (*Oreochromis niloticus*) fingerlings (1.1 g) were fed with water hyacinth treated with various processes. Fresh dry hyacinth (FH), molasses-fermented hyacinth (MF), cow rumen content-fermented hyacinth (RF) and yeast-fermented hyacinth (YF) were incorporated into nine isonitrogenous (35 percent CP), isocaloric (450 kcal GE/100 g) test diets, as a substitute for wheat bran at 10 and 20 percent levels. Fish fed the control diet (wheat bran based) exhibited growth, feed conversion efficiency and production values significantly ($P < 0.05$) higher than those fed with water hyacinth based diets. There was no significant difference in the performance of fish between the fermentation products added at the 10 percent inclusion level. At the 20 percent inclusion level, the performance of fish was further reduced. Despite this rather discouraging result, it is interesting to note that significantly lower growth rate and feed utilization efficiency was found in fish fed with fresh dry water hyacinth than when fish were fed the fermented water hyacinth treatments.

4.5.4 Other processed forms

There are other processing techniques that are employed to increase the feeding value of water hyacinth for livestock, such as boiling, mashing and chopping.

For example, Gohl (1981) reported that boiled water hyacinth is used in Southeast Asia for feeding to pigs. The plants are chopped, sometimes mixed with other vegetable wastes such as banana stems, and boiled slowly for a few hours until the ingredients turn into a paste, to which oil cake, rice bran and sometimes maize and salt are added. The cooked mixture is good for only three days, after which it turns sour. A common formula is 40 kg of water hyacinth, 15 kg of rice bran, 2.5 kg of fishmeal and 5 kg of coconut meal.

In China, mashed water hyacinth is used as feed for Chinese carps (Z. Xiaowei, *pers. com.* 2003). Fresh water hyacinth is mashed into a liquid form with a high-speed beater and applied to ponds for carp fingerlings. The mesophyll cells are considered, rightly or wrongly, similar to phytoplankton. There is an additional means of using mashed water hyacinth as fish feed: water hyacinth pastes are mixed with rice bran and are fermented before applying to the pond.

Kumar *et al.* (1991) evaluated the nutritive value of mashed water hyacinth leaf for rohu spawn (1.9 mg). Mashed water hyacinth leaves were fed in the form of leaf extract. The hyacinth extract was prepared by crushing the leaves with water (1:5) in a heavy-duty mixer. The solution was sieved through a 1 mm mesh to remove the fibrous material. One or two percent common salt was added to the solution. The experiment was conducted for 30 days in 40 L glass tanks. Plankton dominated by rotifers and cladocerans were used as a control treatment. Hyacinth extract was provided at 100 ml/day to the experimental tank containing 120 spawn. The specific growth rate of rohu spawn fed with mesophyll cells was 8.59 while that for the control was 9.04.

Edwards (1987) reported the efficiency of three processing techniques applied to water hyacinth for use as fish feed and fertilizer from unpublished research studies carried out in the Asian Institute of Technology, Thailand. Water hyacinths were added to a series of earthen ponds stocked with *O. niloticus* in three forms: fresh whole plants that decomposed beneath the water in situ; freshly chopped water hyacinth spread on the surface; and composted water hyacinth. Extrapolated yields of 5 to 6 tonnes/ha/year were obtained with all three treatments at the same dry matter loading rate of 200 kg/ha/day (about 3 kg TKN/ha/day).

Bolenz, Omran and Gierschner (1990) suggested the following treatment to avoid the problem of oxalate crystals (see section 4.5.1). The plants should be chopped into small pieces; this helps to eliminate trapped air and negate its ability to absorb water. Then the solid material should be separated from the soluble components in the juice by pressing and centrifugation. The solid phase will be washed with acid to remove the acid-soluble calcium oxalate. The juice may be concentrated, dried and used as a protein source. However, such elaborate treatments will probably not be cost-effective in preparing fish feeds.

4.5.5 Food conversion efficiency

Food conversion values of diets containing varying inclusion levels of dried water hyacinth meal tested for different fish species were included in Table 4.3. It can be observed that the FCR of these test diets varied between >2.0 and <4.0, with the exception of Nile tilapia (FCR 4.3), as reported by Klinavee, Tansakul and Promkuntong (1990) and *Brycon* sp. (FCR 1.7-1.8), as reported by Saint-Paul, Werder and Teixeira (1981). However, it is difficult to standardize an FCR from the available data because of the difference in fish species, water hyacinth inclusion levels, rearing systems and length of rearing. A summary of food conversion ratios for various fish species fed test diets containing fresh and processed water hyacinth is presented in Table 4.4. Apart from the pelleted diets containing dried water hyacinth meal, not much information on FCR for other forms of water hyacinth is available. However, what is available indicates that an FCR value of 3.0 is a reasonably acceptable level for fresh or processed water hyacinths.

TABLE 4.4
Food conversion ratio of fresh and processed water hyacinth for selected fish species

Form of water hyacinth	Incorporation level (%)	Fish species	Fish size (g)	FCR (DM basis)	Reference
Fresh	100	Grass carp	38-104	1.54	Riechert and Trede (1977)
Dried meal as pellet	Various (see Table 4.3)	Various (see Table 4.3)	Various (see Table 4.3)	1.7-4.3	See Table 4.3
Composted	25-75	Nile tilapia	14.2-17.9	2.18-2.57	Edwards, Kamal and Wee (1985)
Decomposed	100	Indian major carps		3.34	Mishra, Sahu and Pani (1988)
Fermented	100	Silver carp and mrigal		2.02-3.72	Olah, Ayyappan and Purushothaman (1990)
Fermented with molasses	20	Nile tilapia	1.1	1.6	El-Sayed (2003)

4.5.6 Digestibility coefficients

Several authors have reported the apparent digestibility coefficients of water hyacinth when fed to carps and tilapia. These varied between species (Table 4.5) and were influenced by the level of water hyacinth incorporation (Table 4.6). Lin and Chen (1983, cited by Wee, 1991) noted that protein from water hyacinth was poorly digested (58.9 percent) by grass carp. Similarly, Riechert and Trede (1977) concluded from their feeding trial with fresh water hyacinths that only 50-60 percent of the feed consumed were actually digested by the grass carp. Apparent protein digestibility (APD) of water hyacinths by Nile tilapia was reported by Pongri (1986, cited by Wee, 1991). He reported APD values of 49-65 percent and 46-65 percent for dried and composted water hyacinth when 37.5 percent of water hyacinth was incorporated in the diet. APD values of water hyacinth leaf meal for Indian major carps (rohu and catla fingerlings) were reported by Hasan and Roy (1994) and Nandeesha *et al.* (1991), respectively (Table 4.6). Digestibility coefficients decreased with increased dietary incorporation of water hyacinth. For rohu, APD values were 65 and 78 percent for 60 and 30 percent incorporation levels, while for catla it varied between 48 and 74 percent at incorporation levels from 45-15 percent. In nature, rohu fingerlings feed predominantly on vegetable debris and microscopic plants while catla are predominantly zooplankton feeders. Therefore, it is likely that rohu would be able to digest plant materials better than catla. In an earlier study with rohu fry (mean weight 0.2 g), Hasan, Moniruzzaman and Omar Farooque (1990) reported the APD of water hyacinth leaf meal to be 55 and 60 percent for 54 and 27 percent levels of dietary water hyacinth inclusion levels. In contrast to these results, Ray and Das (1994) reported much higher APD value (94.0 percent) of water hyacinth leaf meal for rohu fry (3.6 g). Similarly high APD values of water hyacinth for grass carp and common carp fry were reported by Murthy and Devaraj (1990) (Table 4.5).

From the foregoing discussion, it is difficult to draw any definite borderline between digestibility coefficients of carps and tilapia. However, it is apparent that digestibility coefficients are mainly dependent on the level of dietary incorporation. For all practical purposes, the protein digestibility of water hyacinth may safely be taken as 70-80 percent at 15-30 percent dietary incorporation levels, while it may be around 50-60 percent at incorporation levels of 45 percent or above.

TABLE 4.5

Summary of apparent nutrient digestibility coefficients of water hyacinth for selected fish species

Form of water Hyacinth	Fish species	Fish size (g)	Digestibility coefficient (%)			Reference
			Dry matter	Protein	Lipid	
n.s.	Grass carp			58.9		Lin and Chen (1983, cited by Wee, 1991)
Fresh	Grass carp	20-50	50-60			Riechert and Trede (1977)
Dried	Nile tilapia			49-65		Pongri (1986, cited by Wee, 1991)
Composted	Nile tilapia			46-65		Hertrampf and Piedad-Pascual (2000)
Leaf meal	Rohu	3.5		65-78		Hasan and Roy (1994)
Leaf meal	Rohu	0.2		55-60		Hasan, Moniruzzaman and Omar Farooque (1990)
Leaf meal	Rohu	3.6		94	86	Ray and Das (1994)
Leaf meal	Catla	23-32		48-74	63-84	Nandeesha et al. (1991)
Leaf meal	Grass carp	6.5		89	97	Murthy and Devaraj (1990)
Leaf meal	Common carp	3.1		83	98	Murthy and Devaraj (1990)

TABLE 4.6

Apparent nutrient digestibility coefficients of water hyacinth leaf meal for two carp species at different dietary incorporation levels

Fish species	Size (g)	Incorporation level of total diet (%)							Reference
		15	30	45	60	15	30	45	
		Apparent nutrient digestibility (%)							
		Protein				Fat			
Rohu	3.5	-	77.6	-	64.5				Hasan and Roy (1994)
Catla	23-32	73.8	59.9	47.9	-	83.9	77.9	63.1	Nandeesha et al. (1991)

5. Floating aquatic macrophytes – Others

Water hyacinth, duckweed and *Azolla*, the most common and important floating macrophytes, have been described in sections 2, 3 and 4. This section covers other miscellaneous floating macrophytes.

5.1 CLASSIFICATION
A list of commonly occurring species is presented in Table 5.1.

TABLE 5.1
Common and scientific names of some commonly occurring floating aquatic macrophytes

Scientific name	Family	Common names
Commelina bengalensis	Commelinaceae	Pakplab; day flower
Enhydra fluctuans	Compositae	Hinche sak
Enhydra sp.	Compositae	-
Hydrocharis dubia	Hydrocharitaceae	-
Ipomoea aquatica (reptans)	Convolvulaceae	Water spinach
Pistia stratiotes	Araceae	Water lettuce
Salvinia auriculata	Salviniaceae	Water fern
Salvinia cucullata	Salviniaceae	Water fern
Salvinia molesta	Salviniaceae	Water fern
Salvinia natans	Salviniaceae	Water fern
Salvinia rotundifolia	Salviniaceae	Water fern
Trapa bipinosa	Trapaceae	Water chestnut
Trapa natans	Trapaceae	Water chestnut

5.2 CHARACTERISTICS
Like most other macrophytes, these are self-growing plants that are commonly found in the shallow stagnant waters of tropical and sub-tropical countries. The most commonly found are water spinach (*Ipomoea*), water lettuce (*Pistia*), water fern (*Salvinia* spp.) and water chestnut (*Trapa* spp.).

Water spinach (*Ipomoea aquatica*) is a floating plant that roots in marshy soil. It is native to India, SE Asia, and S. China and is commonly eaten as a vegetable (Edwards, 1980).

Water lettuce (*Pistia stratiotes*) is a free-floating aquatic plant that is found growing abundantly in tropical and subtropical region of the world. This plant is very common in lakes, ponds, ditches, irrigation canals, etc. It is reported to be used as a vegetable in India (Varshney and Singh, 1976).

Water fern (Salvinia spp.) is a perennial free-floating aquatic herb belonging to the family Salviniaceae. It has five commonly found species (Table 5.1) and has a wide native range in the neotropics extending from Mexico and the Galapagos Islands through Central America and most of tropical South America as far as southern Brazil (Sculthorpe, 1971). In the Indian sub-continent, the plant was introduced through a route from Brazil to Germany in 1933 (Hadiuzzaman and Khondker, 1993). Reddy and DeBusk (1985) reported the biomass yield of S. rotundifolia to be 32 tonnes DM/ha/year in nutrient non-limiting waters in central Florida, USA. Giant salvinia

(*Salvinia molesta*) is a free-floating clonal fern and reproduces only vegetatively. Though the plant can survive a wide range of temperatures (-3 to 43 °C), optimal growth occurs at 24 to 28 °C (Mcintosh, King and Fitzsimmons, 2003). Stands of salvinia double in 2.2 days with adequate supply of adequate nutrients. During periods of high growth, leaf size decreases and both leaves and stems fold, doubling and layering to cover more of the water surface. This thick plant growth is harmful for other organisms as it prevents light penetration, reduces gaseous exchange and increases biological oxygen demand.

FIGURE 5.1
Water chestnut plants grown in a floodplain, Rangpur, Bangladesh

Water chestnuts (*Trapa* sp.) (Figures 5.1 and 5.2) are floating annual aquatic plants that grow in slow-moving water up to 5 meters deep and are native to warm temperate parts of Asia and Africa. The nut (kernel) of water chestnuts is eaten by humans in raw or cooked form.

5.3 PRODUCTION

In South and Southeast Asia, water spinach is often grown for use as a vegetable in India, Bangladesh, Hong Kong (SAR China), Cambodia, Thailand, Indonesia and Viet Nam, sometimes in ponds fertilized with sewage (Figure 5.3). In China, water lettuce has been reported to be cultivated with two other aquatic macrophytes, water hyacinth and alligator weed (*Alternathera philoxeroides*) (Edwards, 1987). The plants are usually cultivated in rivulets, small bays, or swamps to avoid taking up cultivable land and are usually fed to pigs. Reddy and DeBusk (1984) reported the biomass yield of water lettuce to be 72 tonnes DM/ha/year in nutrient non-limiting waters in central Florida, USA. Water chestnuts are grown in India, China, Indonesia and Bangladesh.

FIGURE 5.2
Water chestnut fruits harvested from a floodplain, Rangpur, Bangladesh

5.4 CHEMICAL COMPOSITION

The chemical composition of seven floating aquatic macrophytes is presented in Table 5.2. Of these the moisture content varied between 85-94 percent. Water spinach had the highest crude protein varying between 24-34 percent, higher lipid contents (2.7-3.9 percent) and low ash (~13 percent) and crude fibre (10.2-12.7 percent) contents. The other two macrophytes were moderately rich in crude protein (12-20 percent), high ash (18-26 percent) and crude fibre (12-20 percent) contents. Clearly, water spinach is nutritionally superior to other macrophytes; its crude protein is comparable

TABLE 5.2
Chemical analyses of some common floating aquatic macrophytes

Floating macrophytes	Moisture (%)	Proximate composition[1] (% DM)					Minerals (% DM)		Reference
		CP	EE	Ash	CF	NFE	Ca	P	
Water fern (*Salvinia cucullata*), leaf, India		11.6	2.8	18.3	20.4	46.9[2]	n.s.	n.s.	Ray and Das (1994)
Water fern (*Salvinia molesta*), leaf, India	77.2[2]	16.3	1.1	21.9	18.5	42.2[3]	n.s.	n.s.	Murthy and Devaraj (1991a)
Water lettuce (*Pistia stratiotes*), leaf, India	93.6[2]	19.5	1.3	25.7	11.7	41.8[3]	2.35[2]	0.30[2]	Murthy and Devaraj (1991b)
Water lettuce (*Pistia stratiotes*), leaf, India		11.4	2.7	18.0	20.1	47.8	n.s.	n.s.	Ray and Das (1994)
Water spinach (*Ipomoea aquatica*), leaf, Philippines		27.5	3.0	13.2	11.6	44.7	n.s.	n.s.	Borlongan and Coloso (1994)
Water spinach (*Ipomoea aquatica*), leaf, Africa	85.0	24.0	2.7	13.3	12.7	47.3	1.20	0.28	Gohl (1981)
Water spinach (*Ipomoea aquatica*), leaf, Fiji	90.8	34.3	3.9	12.9	10.2	38.7	n.s.	n.s.	Gohl (1981)

[1] CP = crude protein; EE = ether extract; CF = crude fibre; NFE = nitrogen free extract; Ca = calcium; P = phosphorus
[2] Data obtained from Tacon (1987)
[3] Adjusted or calculated; not as cited in original publication

FIGURE 5.3
Production of water spinach from Beung Cheung Ek waste water lake, Phnom Penh, Cambodia

These aquatic macropyte are used for both human and/or livestock consumption depending on the season and quality of the plants

Courtesy of William Leschen

only to duckweed. Some information on the amino acid content of various aquatic macrophytes is contained in Annex 1.

5.5 USE AS AQUAFEED

Little work has been conducted on the use of these miscellaneous floating macrophytes as feed for fish.

A summary of selected experimental studies on the use of dried leaf meal of these macrophytes in pelleted diets for various fish species is presented in Table 5.3. Dried water spinach leaf meal has been evaluated as a dietary ingredient/protein source for milkfish and water lettuce (Figure 5.4) and water fern for grass carp, common carp and rohu. The performance of test diets containing various inclusion levels of these macrophyte meals were compared with control diets. The latter comprised fishmeal-based pellets, the traditionally used rice bran-oil cake mixture, or a mixture of fishmeal, oil cake and cereal by-products.

Apart from rohu, these floating aquaphyte leaf meals (Table 5.3) produced reasonably good growth (SGR 2.35-3.29 percent) and food conversion (FCR 1.50-2.11). The performances of milkfish, grass carp and common carp fed various test diets were slightly better or comparable to those obtained for the control diets. However, it should be pointed out that Murthy and Devaraj (1991a, 1991b) conducted their feeding

FIGURE 5.4
A feeding square filled with mixture of grass and chopped water lettuce offered to the fish as supplemental feed (Mymensingh, Bangladesh)

trials with grass carp and common carp in a static experimental setup consisting of a 20 m² cement cistern. A soil bed of 15 cm was provided to each cistern and an initial manuring of cow dung (15 kg per cistern) was carried out prior to the start of the experiment. These authors recorded the plankton volume fluctuating between 0.007 to 0.041 ml/L in the cisterns, indicating that test fish obtained a part of their nutrition from the plankton. The authors noted that the initial manuring produced the plankton at first but the faecal matter of fish and unconsumed food might have been responsible for the continued plankton production. Borlongan and Colso (1994) obtained an SGR of 3.07 percent for milkfish fry with water spinach leaf meal in a growth trial conducted in a recirculatory rearing system. However, in this trial, only 15 percent of the total dietary protein (i.e. 6 percent protein in a 40 percent protein diet) was replaced by water spinach leaf meal. For rohu, Ray and Das (1994) obtained very low SGRs (0.50-0.57 percent) with test diets containing various inclusion levels of water fern in a growth trial conducted in an indoor flow-through rearing system (Table 5.3).when compared with a control diet.

TABLE 5.3
Performance of different fish species to pelleted feeds containing dried floating aquatic macrophytes

Macrophytes/ Fish species	Rearing system	Rearing days	Control diet	Composition of test diet	Inclusion level (percent)	Fish size (g)	SGR (percent)	SGR as percent of control	FCR	References
Water spinach/ milkfish	Brackish water recirculatory system	72	Fishmeal based diet (40 percent protein)	15 percent of total dietary protein from fishmeal replaced by water spinach meal	23	0.3	3.07	100.3	1.50	Borlongan and Coloso (1994)
Water lettuce/ grass carp	Static water outdoor cement cistern	112	Rice bran: groundnut cake (50:50) (27.9 percent protein)	Water lettuce leaf meal, groundnut cake, rice bran and fishmeal (50:22:14:10)	50	6.5	2.52	106.3	1.84	Murthy and Devaraj (1991b)
Water lettuce/ common carp	Static water outdoor cement cistern	112	Rice bran: groundnut cake (50:50) (27.9 percent protein)	Water lettuce leaf meal, groundnut cake, rice bran and fishmeal (50:22:14:10)	50	3.1	3.48	104.5	1.84	Murthy and Devaraj (1991b)
Water lettuce/ rohu	Indoor flow-through system	70	Fishmeal: mustard cake: rice bran (50:35:15) (35 percent protein)	Water lettuce meal incorporated at four levels in control diet by simple displacement.	15　30　45　60	10.0	0.57　0.54　0.54　0.50	107.5　101.9　101.9　94.3	2.00　2.02　2.08　2.66	Ray and Das (1994)
Water fern[1]/ Grass carp	Static water outdoor cement cistern	112	Rice bran: groundnut cake (50:50) (28.0 percent Protein)	Water fern leaf meal, groundnut cake, rice bran and fishmeal (50:24:11:11)	50	6.5	2.35	99.2	2.11	Murthy and Devaraj (1991a)
Water fern[1]/ Common carp	Static water outdoor cement cistern	112	Rice bran: groundnut cake (50:50) (28.0 percent Protein)	Water fern leaf meal, groundnut cake, rice bran and fishmeal (50:24:11:11)	50	3.1	3.29	110.4	2.11	Murthy and Devaraj (1991a)

[1] *Salvinia molesta*
[2] *Salvinia sp.*

Generally, macrophytophagous fish do not consume these floating macrophytes readily in fresh form. Water lettuce is one of the least favoured floating macrophytes for herbivores (Singh *et al.*, 1967). Some scattered reports, however, are available on the consumption of these macrophytes in fresh form. For example, water lettuce is used for feeding Chinese carps in China (Z. Xiaowei, *pers. com.* 2003). Fresh water lettuce is mashed into liquid form with a high-speed beater and is applied to the pond for carp fingerlings. Alternatively, mashed water lettuce is mixed with rice bran and fermented overnight and applied to the pond. Ling (1967, cited by Edwards, 1987) also reported that water lettuce was chopped into small pieces and used to feed grass carp and common carp in China. The plant was also reported to be processed, either mechanically (soaking, mixing, cutting, or grinding) or biologically. The latter involved green storage and fermentation in ditches, tubs or barrels under anaerobic conditions at 65-75 percent moisture after cutting into 6 cm strips and sealing by a 15 cm layer of dry grass topped by a 15 cm layer of moist soil; if the material was too moist it could be sum-dried or mixed with dry hay before sealing. Another floating macrophyte (*Hydrocharis dubia*) is considered a good feed for grass carp in China (Z. Xiaowei, *pers. com.* 2003) and is collected or cultivated for feeding carp fingerlings. In oxbow lakes located in southwestern Bangladesh, water spinach has been found to be one of the preferred macrophytes for grass carp. In the Mymensingh region of Bangladesh, fish farmers often give fresh chopped water lettuce and banana leaves as feed for grass carp and Java barb (*Barbonymus gonionotus*) in their fish ponds. Both of these herbivores readily consume these macrophytes and good results have been observed.

Preliminary studies by Mcintosh, King and Fitzsimmons (2003) involved feeding three month old (3.5 g) Nile tilapia (*Oreochromis niloticus*) with live giant salvinia (*Salvinia molesta*) in a recirculating system, either alone or with a commercial tilapia feed, compared to a control with commercial feed alone. Results indicated that though salvinia was consumed by the fish, there was a weight loss during the 23 day culture period in fish fed salvinia alone. However, some benefit seemed to have been obtained when salvinia was fed in addition to the commercial diet and these authors speculated that a longer feeding period might have demonstrated significant effect. These authors also noted that incorporating dried salvinia as an ingredient in a mixed feed might have produced a more pronounced effect. The primary interest of this research was the use of tilapia in weed control ('biocontrol').

TABLE 5.4
Food conversion ratios (FCR) of selected floating aquatic macrophytes to fish

Macrophytes	Fish species	Fish size (g)	Food conversion ratio		References
			Dry weight basis	Fresh weight basis	
Water spinach	Milkfish	0.3	1.50		Borlongan and Coloso (1994)
Water lettuce	Grass carp			280	Scott and Orr (1970)
Water lettuce	Grass carp			50	De Silva (1995)
Water lettuce	Grass carp	6.5	1.84		Murthy and Devaraj (1991b)
Water lettuce	Common carp	3.1	1.84		Murthy and Devaraj (1991b)
Water lettuce	Rohu	10.0	2.00-2.66		Ray and Das (1994)
Water fern[1]	Grass carp	6.5	2.11		Murthy and Devaraj (1991a)
Water fern[1]	Common carp	3.1	2.11		Murthy and Devaraj (1991a)

[1] *Salvinia molesta*

The food conversion values of diets containing varying inclusion of dried water spinach, water lettuce and water fern (*S. molesta*) tested for various fish species are summarised in Table 5.4. The FCR of water spinach for milkfish was 1.50. However, the inclusion level of water spinach meal was only 23 percent and it remains to be investigated how this might change with increased inclusion levels. The FCR of water lettuce for grass carp and common carp was 1.84 but it varied between 2.00 and 2.66

for rohu, depending on the level of inclusion. The FCR of water fern for grass and common carp was 2.11. For both grass and common carps, the level of inclusion of water lettuce and water fern was 50 percent of the diet.

TABLE 5.5
Digestibility of selected floating and emergent aquatic macrophytes

Macrophytes	Fish species	Fish size (g)	Digestibility (percent)[1]					References
			DM	CP	EE	NFE	CF	GE	
Water lettuce	Rohu	3.6	-	91.5	85.4	92.9	-	52.2	Ray and Das (1994)
Water fern[2]	Rohu	3.6	-	91.4	21.4	66.0	-	64.3	Ray and Das (1994)

[1] DM = dry matter; CP = crude protein; EE = ether extract; CF = crude fibre; NFE = nitrogen free extract; GE = gross energy
[2] *Salvinia cucullata*

It is difficult to standardize the FCR values of these floating macrophytes based on the available data. For example, the inclusion level of water spinach meal was only 23 percent in a fishmeal based pelleted diet, while it is apparent that the good FCR values of water lettuce and water fern obtained for grass and common carps were partially obtained because of the plankton produced in the cement cistern. Nevertheless, FCRs of 2.00-2.50 may be reasonably accepted values for dried water lettuce and water fern when these are used at dietary inclusion levels of up to 50 percent.

Digestibility coefficients of water lettuce and water fern are available only for rohu. Water lettuce was well digested by rohu with digestibility coefficients varied between 85.4 and 92.9 percent for crude protein, crude lipid and NFE (Table 5.5). For water fern, crude protein was well digested (91.4 percent) by rohu but the lipid was poorly digested (21.4 percent). The digestibility of NFE was moderately good (66.0 percent).

6. Submerged aquatic macrophytes

Submerged aquatic macrophytes are defined as plants that are usually rooted in the bottom soil with the vegetative parts predominantly submerged. Many different types of submerged aquatic macrophytes have been identified globally.

6.1 CLASSIFICATION

Most submerged aquatic macrophytes belong to the families Ceratophyllaceae, Haloragaceae, Hydrocharitaceae, Nymphaeaceae and Potamogetonaceae. A list of the most commonly occurring ones is presented in Table 6.1. These macrophytes are found in various types of water bodies, including estuaries, rivers, lakes, ponds, natural depressions, ditches, swamps and floodplains. Like other macrophytes, they compete with phytoplankton for nutrients, decreasing the productivity of the water and causing hindrance to the movement of fish, irrigation and navigation.

6.2 CHARACTERISTICS

Submerged macrophytes are distributed all over the world except some very deep and cold water lakes in polar countries. For example, submerged macrophytes are found in Asia, mid-eastern Europe, eastern Africa, north and Central America and Australia and thus have a wide range of environmental requirements in their natural habitats. Submerged aquatic macrophytes are more commonly found in shallow stagnant waters. Some comments on environmental effects on the composition of submerged aquatic macrophytes are given in section 6.4.

TABLE 6.1

Common and scientific names of various submerged aquatic macrophytes used as fish feed

Scientific name	Family	Common name
Blyxa lancifolia	Hydrocharitaceae	Blyxa
Cabomba caroliniana	Nymphaeaceae	Fanwort
Ceratophyllum demersum/ C. submersum	Ceratophyllaceae	Hornwort/Coontail
Chara sp.	Characeae	Chara
Elodea canadensis	Hydrocharitaceae	Canadian pondweed
E. densa	Hydrocharitaceae	Brazilian pondweed
E. trifoliate	Hydrocharitaceae	Pondweed
Haterrauthera limosa	Hydrocharitaceae	Water stargrass
Hydrilla verticillata	Hydrocharitaceae	Oxygen weed
Myriophyllum aquaticum	Haloragaceae	Water milfoil
M. exalbescens	Haloragaceae	Water milfoil
M. spicatum	Haloragaceae	Eurasian water milfoil
Najas graminea	Hydrocharitaceae	Water velvet/ Najas
N. guadalupensis	Hydrocharitaceae	Water velvet/ Najas
N. marina	Hydrocharitaceae	Water velvet/ Najas
Ottelia alismoids	Hydrocharitaceae	Ottelia
Potamogeton crispus	Potamogetonaceae	Curlyleaf pondweed
P. gramineous	Potamogetonaceae	Pondweed
P. nodosus	Potamogetonaceae	Longleaf pondweed
P. pectinatus	Potamogetonaceae	Sago pondweed
Ruppia maritima	Potamogetonaceae	Ruppia
Utricularia vulgaria	Nymphaeaceae	Bladderwort
Vallisneria Americana	Hydrocharitaceae	Eelgrass
V. spiralis	Hydrocharitaceae	Eelgrass

6.3 PRODUCTION

Production or cultivation techniques have not been developed for most of the submerged macrophytes, probably because this has not been necessary. However, some are used as human food and are therefore cultivated. The tip of the shoots of the Eurasian water milfoil (*Myriophyllum spicatum*) is eaten as a vegetable in Java (Indonesia) and is cultivated there (Cook *et al.*, 1974). The leaves of *Blyxa lancifolia* are eaten as vegetables in India, where it is one of the most popular vegetables and is eaten raw with certain kinds of fish. Another submerged plant, *Ottelia alismoides*, is also used for human consumption. The entire plant, except the roots, is cooked as a vegetable. Information on the standing crop of submerged macrophytes is scarce, except that Boyd (1968) reported that the standing crop value of submerged plants and algae in lakes in Alabama ranged from 1-4 tonnes/ha. Westlake (1966) reported net production of submerged macrophytes ranging from 4 to 20 tonnes DM organic matter/ha/year in fertile ponds.

6.4 CHEMICAL COMPOSITION

Chemical analyses of some of the common submerged macrophytes used as fish feed are presented in Table 6.2. Submerged macrophytes generally have a high water content, which is usually a major deterrent to their harvest and utilization (Edwards, 1980). The water content of the submerged macrophytes listed varied from 84 to 96 percent. The water content of hornwort (*Ceratophyllum demersum*) is particularly high (93-96 percent) and it can thus be described as an 'absolutely succulent' type of macrophyte. The crude protein values of these macrophytes varied between 9 and 22 percent DM, although most contained levels of 13-15 percent. Most of the submerged plants contained less than 4 percent lipid, although there were some exceptions, particularly for oxygen weed. The ash content varied widely from 10 to over 56 percent; however, most values were between 15 and 30 percent. Fibre contents varied from 7 to 37 percent but values between 7 and 11 percent were more common.

The apparently wide variations in proximate composition are due to both interspecific and intraspecific differences in macrophytes. For example, Boyd (1968) reported crude protein and ash contents of 10.9 and 16.0 percent respectively for curlyleaf pondweed (*Potamogeton crispus*), whereas Pine, Anderson and Hung (1990) reported values of 15.2 and 49.2 percent respectively for crude protein and ash for the same macrophyte. Similarly, considerable intraspecific variations in nutritional composition in hornwort, long leaf pondweed (*P. nodosus*), oxygen weed (*Hydrilla verticillata*) and water velvet (*Najas guadalupensis*) were observed by these authors. These variations were more pronounced in the case of ash and fibre contents than protein and lipid. Such intraspecific variations in nutritional composition may also be attributed to variations in geographic locations, seasonality and environment.

Muztar, Slinger and Burton (1978) recorded a large variation in crude protein content (7.5-14.9 percent) in Eurasian water milfoil (*M. spicatum*), simply due to difference of locations and seasons, although the plant samples were collected from the same lake in Canada. There is evidence that the crude protein content increases as the nutrient content of the water in which the plant is grown increases. Pine, Anderson and Hung (1989) recorded marked variations in proximate composition and acid detergent fibre of three macrophytes species (sago pondweed *P. pectinatus*, long leaf pondweed *P. nodosus* and Eurasian water milfoil) grown in canals with either static or flowing water. The greatest differences found were in the levels of dry matter (DM), nitrogen-free extract, ash, and acid detergent fibre. These major variations in proximate composition were possibly correlated with the morphological forms that the plants developed as a response to either static or flowing water conditions. Larger shoots were produced in these three macrophytes when grown in canals with flowing water as opposed to static water (Pine, Anderson and Hung, 1989). Furthermore, Pine,

TABLE 6.2
Chemical analyses of some common submerged aquatic macrophytes

	Moisture (percent)	Proximate composition[1] (percent DM)					Minerals[1] (percent DM)		Reference
		CP	EE	Ash	CF	NFE	Ca	P	
Brazilian pondweed (*Elodea densa*), USA	90.2	20.5	3.3	22.1	29.2[3]		n.s.	n.s.	Boyd (1968)
Canadian pondweed (*E. canadensis*), USA	91.1[2]	14.1	1.4	56.5	14.7[2]		1.75[2]	0.36[2]	Pine, Anderson and Hung (1990)
Chara (*Chara sp.*), fresh, USA	91.6	17.5	1.6	35.8	23.8[3]		n.s.	n.s.	Boyd (1968)
Chara (*C. vulgaris*)	91.6	8.8	0.8	28.3	14.0	48.1[5]	n.s.	n.s.	Tacon (1987)
Curlyleaf pondweed (*Potamogeton crispus*), USA	88.2	10.9	2.9	16.0	37.2[3]		1.68[2]	0.24[2]	Boyd (1968)
Curlyleaf pondweed (*P. crispus*), USA[4]		15.2	1.3	49.2			n.s.	n.s.	Pine, Anderson and Hung (1990)
Eurasian water milfoil (*Myriophyllum spicatum*), USA[4]	86.1	21.8	1.7	25.0	7.5	44.0[5]	2.82[2]	0.41[2]	Pine, Anderson and Hung (1989)
Fanwort (*Cabomba caroliniana*), USA	93.0	13.1	5.4	9.6	26.8[3]		n.s.	n.s.	Boyd (1968)
Hornwort (*Ceratophyllum demersum*), India	93.2	13.7	3.1	30.5	7.5	45.2[5]	1.30[2]	0.32[2]	Venkatesh and Shetty (1978b)
Hornwort (*C. demersum*), India	95.8	12.9	2.6	32.4	9.1	43.0	n.s.	n.s.	Hajra (1987)
Hornwort (*C. demersum*), Thailand		16.2	1.5	19.7	8.3	54.3[5]	n.s.	n.s.	Chiayvareesajja et al. (1988)
Hornwort (*C. demersum*), USA	94.8	21.7	6.0	20.6			n.s.	n.s.	Boyd (1968)
Long leaf pondweed (*P. nodosus*), USA	84.2	11.2	3.6	10.9	21.7[3]		n.s.	n.s.	Boyd (1968)
Long leaf pondweed (*P. nodosus*), USA[4]	92.7	14.6	1.5	45.2	6.5	32.2	n.s.	n.s.	Pine, Anderson and Hung (1989)
Oxygen weed (*H. verticillata*), India	89.8	14.6	7.3	21.6	11.1	45.4[5]	4.40[2]	0.28[2]	Venkatesh and Shetty (1978b)
Oxygen weed (*H. verticillata*), leaf, India		14.1	6.5	19.3	6.9	53.2[5]	n.s.	n.s.	Ray and Das (1994)
Oxygen weed (*H. verticillata*), leaf, Sri Lanka	92.3	21.5	11.3				n.s.	n.s.	De Silva and Perera (1983)
Sago pondweed (*P. pectinatus*), USA[4]	91.8	14.6	1.8	40.8	7.7	35.1[5]	n.s.	n.s.	Pine, Anderson and Hung (1989)
Water velvet (*Najas guadalupensis*)		15.7	11.6	10.6	27.8	34.3[5]	0.98[2]	0.15[2]	Buddington (1979)
Water velvet (*N. guadalupensis*), USA	92.7	9.7	3.9	22.7	29.5[3]		n.s.	n.s.	Boyd (1968)

[1] CP = crude protein; EE = ether extract; CF = crude fibre; NFE = nitrogen free extract; Ca = calcium; P = phosphorus
[2] Data from Tacon (1987)
[3] Cellulose
[4] Mean of proximate composition values of weeds collected from flowing and static water
[5] Adjusted or calculated; not as cited in original publication

Anderson and Hung (1990) observed marked differences in the proximate composition of three aquatic macrophyte species (curlyleaf pondweed *P. crispus*, Canadian pondweed *Elodea canadensis* and Eurasian water milfoil) grown in canals having static and flowing water for three seasons (winter, summer and fall). For example, during the winter, the ash content in Eurasian water milfoil was 34.6 percent in static water but 43.5 percent in flowing water. These macrophytes also exhibited significant differences in lipid levels when growing in static or in flowing water, namely 0.5 and 2.0 percent (curlyleaf pondweed), 0.4 and 2.65 (Canadian pondweed), and 0.55 and 1.8 percent (Eurasian milfoil).

For all practical purposes, the crude protein content of Brazilian pondweed and water milfoil may be assumed to be around 20–22 percent, whereas for other submerged macrophytes it may be taken as 13–16 percent (although some exceptions are shown in Table 6.2). Similarly, the crude lipid content of most of the submerged macrophytes is around 4 percent or below, except for fanwort and oxygen weed, which are >5 percent and some individual analyses for hornwort and water velvet (Table 6.2). The extent of intraspecific variation does not permit species-wise generalizations for the ash and fibre contents of submerged macrophytes.

FIGURE 6.1
Grass carp - a voracious macrophyte feeder

A grass carp harvested from a private fish farm in Mymensingh, Bangladesh
Courtesy of M.C. Nandeesha

6.5 USE AS AQUAFEED

A review of the literature indicates that an extensive number of research studies have been carried out on various submerged macrophytes in different parts of the world. However, most of these studies concern effective control of submerged macrophytes by herbivorous fish. Reports are also available on the species preference and consumption rates of submerged aquatic macrophytes by herbivorous fish. Submerged aquatic macrophytes are generally soft in nature, moderately rich in protein and are preferred by different herbivorous fish. In spite of these attractive qualities, only a limited number of research studies have been carried out on their potential utilization as fish feed in pond aquaculture. The results of these studies are variable and species dependent. The most commonly used as fish feed are chara (*Chara* sp.) hornwort, oxygen weed (*Hydrilla*), water velvet (*Najas*), water milfoil (*Myriophyllum*) and pondweeds (*Elodea*). Most studies were on grass carp (Figure 6.1) and tilapia and the submerged macrophytes were fed either in fresh form or as a dried meal within a pelleted diet.

6.5.1 Research studies

A summary of results of selected growth studies carried out on the use of fresh submerged aquatic macrophytes for fish is presented in Table 6.3. Fresh macrophytes are generally given to macrophytophagous fish, either whole or after being cut into small pieces.

In experiments with controlled feeding regimes wherein experimental fish were fed fresh macrophytes as a complete diet in clear water systems (glass aquaria or fibre glass tanks), growth responses were either very poor or negative growth was displayed (Table 6.3). For example, Hajra (1987) reported an SGR of 0.23 percent for grass carp when hornwort was fed *ad libitum* in a clear water fibreglass rearing system.

TABLE 6.3
Performance of tilapia, grass carp and shrimp fed fresh submerged aquatic macrophytes

Macrophytes/ fish species	Rearing system	Rearing days	Control diet	Feeding rate (percent BW/day) Fresh weight	Feeding rate (percent BW/day) Dry weight	Fish size (g)	SGR (percent)	SGR as percent of control	FCR Fresh weight	FCR Dry weight	References
Hornwort/ Nile tilapia	Circular cages in lake	395	Chicken pellet (19.9 percent protein)	n.s.	5	21.7	0.33[1]	42.3	n.s.	15.2	Tantikitti et al. (1988)
Hornwort/ Nile tilapia	Cages in lake	153	Chicken pellet (19.9 percent protein)	n.s.	5	16.7	0.95[2]	69.9	n.s.	4.35	Chiayvareesajja et al. (1988)
Hornwort/ grass carp	Cement cistern	182	-	125	8	12.0	0.94		128.4	10.3	Venkatesh and Shetty (1978a)
Hornwort/ grass carp	Static fibreglass tank	30	-	Ad libitum		14.4	0.23		97.6	4.1	Hajra (1987)
Hornwort/ grass carp	Static fibreglass tank	15	-	Ad libitum		52.2	0.21		96.4	4.05	Hajra (1987)
Oxygen weed/ grass carp	Cement cistern	182	-	100	10	12.0	1.17		94.0	9.4	Venkatesh and Shetty (1978a)
Oxygen weed/ grass carp	Irrigation tank	126	-	Ad libitum		500	0.94		132.0		Keshavanath and Basavaraju (1980)
Oxygen weed/ grass carp	Cement cistern	120	-	Ad libitum		3.0	4.27		45.6		Devaraj, Manissery and Keshavappa (1985)
Chara/ Tilapia zillii	Glass aquaria	56	-	Ad libitum		17.9	0.15		n.s.	n.s.	Saeed and Ziebell (1986)
Najas marina/ Tilapia zillii	Glass aquaria	56	-	Ad libitum		16.7	0.12		n.s.	n.s.	Saeed and Ziebell (1986)
Elodea densa/ Tilapia zillii	Glass aquaria	56	-	Ad libitum		20.5		Negative growth displayed			Saeed and Ziebell (1986)
Myriophyllum exasbescens/ Tilapia zillii	Glass aquaria	56	-	Ad libitum		19.6		Negative growth displayed			Saeed and Ziebell (1986)
N. gramineal shrimp (Penaeus monodon)	Glass tank	30	Commercial pellet (40 percent protein)	Ad libitum		PL$_{50}$ (0.5 g)	5.26	102.0	n.s.	n.s.	Primavera and Gacutan (1989)
Ruppia maritima/ shrimp (Penaeus monodon)	Glass tank	30	Commercial pellet (40 percent protein)	Ad libitum		PL$_{50}$ (0.5 g)	2.54	49.0	n.s.	n.s.	Primavera and Gacutan (1989)

[1] SGR of control diet was 0.78 percent, while fish fed no supplementary diet had SGR of 0.55 percent;
[2] SGR of control diet was 1.36 percent, while fish fed no supplementary diet had SGR of 1.14 percent

Similarly, poor or negative growth responses were recorded when *T. zillii* were fed various submerged macrophytes (*Chara* sp., *N. marina*, *E. dens* and *M. exalbescens*) in a clear water static glass aquarium. Poor performances of Nile tilapia were also recorded by Tantikitti *et al.* (1988) when fed with fresh hornwort in cage culture. These authors evaluated fresh hornwort as feed for Nile tilapia and compared its growth and profitability with chicken pellets and without supplementary feed. In a 14 month trial in Songkhla lake in Thailand, chicken pellets produced the best growth (weight gain 290 g, SGR 0.78 percent/day), while the performances of fish fed fresh hornwort and those not provided with any supplementary feed were similar (hornwort: weight gain 65.25 g, SGR 0.33 percent/day; no supplementary feed: weight gain 87.7 g, SGR 0.55 percent/day). Fish fed fresh hornwort did not have any advantage over fish cultured without any aquatic weed, either in growth or profitability.

Fish reared in clear water static systems tend to consume much less macrophytes than those reared in cement cisterns and ponds/tanks. Hajra (1987) reported a hornwort consumption rate of 25 percent BW/day for grass carp in glass aquaria while the feeding rate used by Venkatesh and Shetty (1978a) for grass carp for the same macrophyte in cement cisterns was 100 percent BW/day. The variability in growth responses between clear water indoor static systems and outdoor rearing systems/ponds might be attributed to the differences in their consumption rates. Moreover, submerged aquatic macrophytes usually contain about 13-16 percent protein (Table 6.2). The dietary protein requirement of tilapia and grass carp is much higher (32-40 percent), which the macrophytes could not generally provide. Therefore, fish cultured only on a macrophyte diet either lose weight or grow very slowly. The better growth responses in cement cisterns, earthen ponds or tanks can also be attributed to the presence of other food organisms such as plankton, benthos, etc.

It is difficult to compare the performances of different macrophytes because of the variability of rearing systems, experimental duration and fish species. Nevertheless, grass carp appeared to have performed better when fed oxygen weed than when fed hornwort (Figure 6.2). Venkatesh and Shetty (1978a) obtained an SGR of 0.94 percent BW/day for hornwort, while an SGR of 1.17 was recorded for oxygen weed in the same experimental study. Devaraj, Manissery and Keshavappa (1985) recorded an SGR of 4.27 percent for grass carp by feeding oxygen weed *ad libitum* in an experimental study conducted for 120 days. CIFA (1981) found hornwort to be a poor inducer of growth, probably due to its poor digestibility.

Attempts have also been made to use dried submerged macrophytes in pelleted feeds for fish. Drying reduces the moisture content and increases the stability and form of macrophytes. However, the number of studies is extremely limited. A summary of the results of growth studies carried out on the use of hornwort meal in dry or semi-moist pelleted feeds for Nile tilapia is presented in Table 6.4. Test diets were prepared by using varying inclusion levels of hornwort meal ranging from 40-98 percent in combination with rice bran and/ or fishmeal. In these studies the performances of fish fed the test diets were sometimes compared with control diets that consisted of chicken pellets or commercial fish pellets containing 16.8-20.7 percent crude

FIGURE 6.2
Farmers carrying mixture of hornwort and oxygen weed in rickshaw van for feeding their fish (Jessore, Bangladesh)

protein. In all cases where control diets were used the performances of fish fed the test diets were significantly lower than the control. In some cases the fish fed the test diets produced growth responses even lower than those given no supplementary feed. For example, Chiayvareesajja *et al.* (1988) fed test diets containing various inclusion levels of hornwort meal and obtained SGRs varying from 1.01 to 1.21 in cages, while the SGR of the control diet was 1.36 percent and the fish given no supplementary feed had a SGR of 1.14. It should also be pointed out that the control diets themselves may have produced sub-optimal growth, as their protein contents varied between 16.8–20.7 percent, much lower than the optimum requirements of grass carp found when a complete diet is tested in a clear water system.

6.5.2 On-farm utilization

Reports on the on-farm utilization of submerged macrophytes are rather limited. Bala and Hasan (1999) reported the efficient on-farm utilization of submerged macrophytes in oxbow lakes located in southwestern Bangladesh. Oxbow lakes (local name: *baors*) are semi-closed water bodies, cut off from old river channels in the delta of the Ganges. There are approximately 600 oxbow lakes in southwestern Bangladesh, with an estimated combined water area of 5 000 ha. Many of these oxbow lakes have been brought under culture-based fisheries management by screening the inlets and outlets.

FIGURE 6.3
Cultivation of watercress, *Nasturtium officinale* in a bamboo frame for feeding of fish in cages (Son La Province, Viet Nam)

Courtesy of M.G. Kibria

Six carp species, i.e. Indian major carps (rohu, catla, mrigal), Chinese carps (silver carp and grass carp) and common carp, are regularly stocked and harvested almost throughout the year. The stocking density and species ratios vary widely between lakes and depend on the water colour and presence of macrophytes in the lake (Hasan and Middendorp, 1998; Bala and Hasan, 1999). Fishers generally stock more silver carp in lakes with green water and more grass carp in lakes with a greater coverage of floating and submerged macrophytes. The most commonly available aquatic macrophytes in oxbow lakes are water hyacinth (*Enhydra fluctuans*), water spinach (*Ipomoea aquatica*), duckweed (*Lemna minor* and *L. major*), oxygen weed, hornwort, pondweeds (*P. crispus* and *P. nodosus*), eelgrass (*Vallisneria spiralis*), monocharia (*Monochoria hastata*), lotus (*Nelumbo nucifera*) and water lily (*Nymphaea* spp.). The most preferred aquatic macrophytes for grass carp in oxbow lakes are water spinach, duckweed, oxygen weed, hornwort and pondweeds (*Potamogeton*). Grass carp also eat the tender leaves of eelgrass. Average stocking densities and yields of each fish species, grouped by the predominant water colour (green, brown and clear) of 14 oxbow lakes managed under the Oxbow Lakes Project II are shown in Table 6.5. Green water lakes are oxbow lakes with distinct algal blooms, as indicated by low Secchi readings, and also generally have little or no aquatic vegetation. On the other hand brown water lakes have comparatively more aquatic vegetation. Clear water lakes mostly have a comparatively high cover of floating and submerged aquatic vegetation. Green water lakes produce the highest yield of silver carp while a higher yield of grass carp is recorded in clear water lakes.

On-farm utilization of aquatic macrophytes in cage culture in oxbow lakes in southwestern Bangladesh (Figure 6.4) has also been observed by the first author of this

TABLE 6.4
Performance of Nile tilapia (*Oreochromis niloticus*) fed pelleted feeds containing dried hornwort (*C. demersum*) meal

Rearing system	Rearing days	Control diet	Composition of test diet	Inclusion level (percent)	Fish size (g)	SGR (percent)	SGR as percent of control	FCR	References
Cages in lake	153	Chicken pellet (19.9 percent protein)	Hornwort meal incorporated at nine levels in combination with pig feed or fishmeal.	47.5-97.7	11.7-18.5	1.01-1.21[1]	74.3-89.0	3.87-4.10	Chiayvareesajja et al. (1988)
Indoor static plywood tank	63	No control diet	HW meal: rice bran: fishmeal (4:3:1)	57.1	3.1	2.27	-	-	Chiayvareesajja and Tansakul (1989)
Earthen pond	153	No control diet	HW meal: rice bran: fishmeal (4:3:1)	57.1	4.4	1.73	-	4.05	Chiayvareesajja et al. (1989)
Cages in lake	90	Commercial pellet (20.7 percent protein)	HW meal: rice bran: fishmeal (4:3:1) HW meal: rice bran: fishmeal (4:2:2)	57.1 50.0	88.0 98.2	0.52 0.79	61.5 93.5	n.s. n.s.	Chiayvareesajja, Wongwit and Tansakul (1990)
Clear water fibre glass tank	77	Chicken pellet (16.8 percent protein)	HW meal, fishmeal and rice bran (40:14.6:40.2)	40.0	7.2	1.09	69.0	3.70	Klinavee, Tansakul and Promkuntong (1990)

[1] SGR of control diet was 1.36 percent, while fish fed no supplementary diet had SGR of 1.14 percent

document. Selected fresh submerged, floating and emergent aquatic macrophytes are used as feed for fingerling rearing in cages floated in oxbow lakes by farmers, with the help of local NGOs. Grass carp, common carp, Java barb (*Barbonymus gonionotus*) and Nile tilapia fry (1.5-2.0 inch) are stocked and reared for about two months until they attained about 4-6 inch. Indian major carps (rohu and mrigal) are also occasionally stocked. The stocking rate varies between 1 400 and 1 600 per 8 m^3 (2 m x 2 m x 2 m) cage. Two stocking combinations are normally used: grass carp, common carp and tilapia; or grass carp and Java barb. Grass carp generally form the bulk (70-75 percent) of the stock. Chopped or whole fresh macrophytes are put into the cages in the morning, along with 3 kg of a rice bran–wheat bran–oil cake mixture (7:1:2). *Ad libitum* feeding or a fixed quantity of 4-5 kg of fresh macrophytes is provided to each cage every day. The most commonly used macrophytes are: submerged – pondweeds, oxygen weed, hornwort and eelgrass; floating – duckweed (*Wolffia arrhiza*); and emergent - *Monochoria hastata*. Pondweeds, oxygen weed, hornwort and duckweed are readily eaten by grass carp, tilapia and Java barb, whereas the roots and tender leaves of *Monochoria* and the tender leaves of eelgrass are generally eaten only by grass carp. Good results are obtained with grass carp and tilapia/Java barb. Jagdish, Rana and Agarwal (1995) and Aravindakshan *et al.* (1999) recommended the use of aquatic macrophytes such as *Hydrilla, Najas, Ceratophyllum* and duckweeds as food for grass carp.

FIGURE 6.4
Mixtures of selected fresh submerged, floating and emergent aquatic macrophytes are given as feed for fingerling rearing in cages floated in oxbow lakes in southwestern Bangladesh

Macrophyte preferences

Soft submerged aquatic macrophytes are readily eaten by certain herbivorous fish. The most commonly fed are hornwort, oxygen weed, water velvet, water milfoil and pondweeds. The most efficient herbivorous fish is probably the grass carp (known in the USA as the white amur). Grass carp feed voraciously on submerged aquatic macrophytes. Several investigations have been carried out to find the consumption rates and preferences of submerged aquatic macrophytes by this herbivorous fish.

Although grass carp are not specialized feeders and have been reported to consume over 170 different species of aquatic macrophytes (Redding and Midlen, 1992), they

TABLE 6.5
Mean stocking densities and yields of six carp species, grouped by the predominant water colour (green, brown and clear) of 14 oxbow lakes managed under Oxbow Lakes Project II

Stock/ Yield	Water colour	Silver carp	Catla	Rohu	Common carp	Mrigal	Grass carp	Total
Stocking density (no/ha)	Green	1 785	387	519	322	616	216	3 845
	Brown	997	325	740	634	296	345	3 337
	Clear	265	197	598	199	247	423	1 929
Yield (kg/ha)	Green	317	76	99	73	77	58	700
	Brown	174	58	101	52	36	64	485
	Clear	25	34	115	33	9.3	86	307

Source: modified from Bala and Hasan (1999)

were shown to have a preference for certain macrophytes over others. Cassani (1981) noted that grass carp prefer submerged, rather than floating macrophytes when they are supplied in fresh form. According to Prabhavathy and Sreenivasan (1977), grass carp are known to ignore all aquatic vegetation in the presence of oxygen weed. Venkatesh and Shetty (1978a, 1978b) fed two submerged aquatic macrophytes (oxygen weed and hornwort) to grass carp and observed that oxygen weed was the most readily consumed, the whole plant being eaten in the process. In the case of hornwort, these authors recorded that the smaller fish preferred only the leaves, while the bigger fish fed readily on the entire plant. In another study, Bhukaswan, Pholprasith and Chatmalai (1981) reported that grass carp preferred submerged macrophytes such as water velvet and oxygen weed and floating macrophytes such as water fern. Mitzner (1978) found that grass carp of approximately 380 g have a preference for *Najas* and *Potamogeton*. The feeding preferences of the blue tilapia *Tilapia aurea* (weight ranging from 94-176 g) for five aquatic plants were tested by Schwartz and Maughan (1984). These authors found that the order of preference was (1) *Najas guadalupensis* and *Chara* sp.; (2) filamentous algae (predominantly *Cladophora* sp.); (3) *Potamogeton pectinatus* L.; and (4) *P. nodosus*.

However, the results of many studies on the preferences of grass carp and their feeding rates are not in agreement. For example, *E. densa*, a non-preferred macrophyte was eaten at the lowest rates in trials in the Pacific Northwest of the USA but proved to be the first choice and eaten rapidly in trials in Florida (Van Dyke. Lestie and Nall, 1984) thus contradicting other findings that this plant was only moderately preferred and consumed. Hornwort was quickly eaten in Arkansas and Colorado lakes, but not in Florida. Similarly, Bonar *et al.* (1990) recorded that grass carp fed on *E. canadensis* from three lakes at significantly different rates, but ate *E. densa* from two of the sites at similar rates. The latter authors further observed that the feeding rate of the grass carp was positively correlated with the concentration of calcium and lignin, but negatively correlated to the content of iron, silica and cellulose, the most important predictors for consumption rate being calcium and cellulose.

Hickling (1966), Prowse (1971) and Wiley, Pescitelli and Wike, (1986) hypothesized that feeding rate and preference in grass carp were primarily influenced by the time it took the fish to process or 'handle' the plant. Its fibre content or the encrustation on its surface can affect the handling time. The coarseness of macrophytes, due to the encrustation by calcium carbonate on their external surfaces, makes them unpalatable (Boyd, 1968). Because grass carp do not digest cellulose, plant cell walls must be masticated before contents can be assimilated (Hickling, 1966). Wiley, Pescitelli and Wike (1986) thought that this would increase the handling time of plants high in cellulose and should lower the preference ranking and the rate of consumption.

Pine, Anderson and Hung (1989) reported the results of a study where triploid grass carp were presented with three submerged aquatic macrophytes species (sago pondweed, Eurasian water milfoil, and longleaf pondweed) in outdoor canals with static and flowing water in winter, spring and summer. During spring and summer, grass carp showed distinct variation in their preference for aquatic weed types, depending on their environmental conditions. Plants of all three species produced longer shoots in canals with flowing water than with static water. The differences in shoot length might have altered the consumption rate and preference of the fish. Flowing conditions also had varying effects on the nutritional content of the plants, as shown in proximate analyses. The preference of triploid grass carp, however, had no correlation with the proximate analysis variables of the macrophytes. This suggests that accessibility and ease of mastication were more important in determining preference than the nutritional quality of the plants. In a further study, Pine, Anderson and Hung (1990) observed significant variations in feeding preferences and feed efficiencies of one year old grass carp for three submerged macrophytes (curlyleaf pondweed, Canadian pondweed and Eurasian water milfoil) depending on the season (winter, summer and fall) and the flow of canal water (static and flowing). These authors attributed the differences in feeding preferences partly to the accessibility of plants to the fish (owing to the difference in plant stature); plants in static canals did not grow as long as those in flowing canals.

Aquatic macrophyte preferences of grass carp have also been found to be affected by the ambient temperature. Redding and Midlen (1992) reported that grass carp consumed more of the softer and more succulent submerged aquatic macrophytes, such as *Elodea*, *Hydrilla*, *Myriophyllum* and *Potamogeton*, when water temperatures were below 12-15 °C.

The discrepancies in the results of the various studies reviewed above suggest that ranking plant palatability on the basis of species type alone would be an oversimplification. Environmental factors and fish size may also play important roles in determining the macrophyte preferences and consumption rates of grass carp.

Other herbivorous fish are known to consume submerged aquatic macrophytes, such as tilapia (*Tilapia zillii* and *T. rendalli*), Java barb (*Barbonymus gonionotus*) and giant gourami (*Osphronemus gorami*). It has also been reported that the silver barb (*Puntius gonionotus*) controlled dense vegetation of *Ceratophyllum* and *Najas* from a 284 ha reservoir in East Java, Indonesia within 8 months of stocking (Schuster, 1952 cited by Edwards, 1980). This author also noted that *T. zillii* and *T. rendalli* are voracious feeders of submerged macrophytes. *T. zillii* feeds on various types but shows preferences when feeding choices are offered. For example, Buddington (1979) reported that *T. zillii* preferred *Najas guadalupensis* as a food source to *Lemna*, *Myriophyllum spicatum* and *Potamogeton pectinatus*. Saeed and Ziebell (1986) conducted an experimental study by feeding different macrophytes (*Chara sp.*, *Najas marina*, *Elodea densa* and *Myriophyllum exalbescens*) to *T. zillii* and observed that the most preferred macrophyte was *Chara* followed by *N. marina*. *E. densa* and *M. exalbescens*. These authors noted that the coarseness of these macrophytes appeared to have some influence on its consumption by the fish. *N. marina* has characteristically sharp-toothed leaf margins. Fish avoided the terminal bushy twigs on which the leaves are crowded while taking stems and lower leaves, probably because the spines are less numerous. Similarly, *T. zillii* avoided the bulky stems of *E. densa* and fed on the leaves and soft slender stems, which are easy to grasp and separate. Like grass carp, *T. zillii* also showed a diet shift with increase in size. *T. zillii* over 9.0 cm long were able to eat macrophytes better than their juveniles. *O. gorami* is another fish that feeds mainly on plant leaves and was introduced into irrigation wells in India from Java to control submerged macrophytes (Edwards, 1980).

Consumption levels
Ad libitum feeding of fresh macrophytes is generally used for herbivorous fish, although fresh weight feeding rates of 100-150 percent of body weight (BW)/day are occasionally recommended for grass carp. These empirical feeding rates have probably been derived from field observations of the consumption rates of different macrophytes by grass carp, as reviewed below. The consumption rates of oxygen weed and hornwort for grass carp were reported to be 100-150 percent BW/day (Singh *et al.*, 1967; Bhatia, 1970). Opuszynski (1972) reported that the consumption rates for smaller sized grass carp were as high as 100-200 percent BW/day. Based on their field observations and calculations, Shireman and Maceina (1981) suggested four empirical consumption rates of grass carp for oxygen weed. These were: 100 percent BW/day for grass carp up to 3 kg; 75 percent BW/day for 3-4 kg; 50 percent BW/day for 4-6 kg; and 25 percent BW/day for >6 kg. Venkatesh and Shetty (1978a, 1978b) used fresh weight feeding rates of 100 percent and 125 percent BW/day for oxygen weed and hornwort respectively, in

a growth trial with grass carp. These authors observed that the these restricted feeding rates might not have been adequate and recommended *ad libitum* feeding for grass carp Bhukaswan, Pholprasith and Chatmalai (1981) reported that grass carp (<1.0 kg) consume water velvet at levels as high as 243 percent and oxygen weed as high as 191 percent BW/day. In contrast, Hajra (1987) reported much lower consumption rates of hornwort by grass carp. The mean daily dry matter intake per 100 g body weight was 0.837 g and 0.977 g in small (14.4 g) and large (52.2 g) fingerlings, respectively. The fresh weight consumption approximated 25 percent of body weight.

Saeed and Ziebell (1986) recorded distinct variation in consumption while feeding four different submerged macrophytes *ad libitum* to *T. zillii*. The consumption rates were 79 percent, 67 percent, 24 percent and 16 percent BW/day for *Chara sp., N. marina, E. densa* and *M. exalbescens* respectively.

Food conversion rates
Food conversion values of diets containing varying inclusion levels of dried hornwort meal in pelleted diets fed to Nile tilapia were presented in Table 6.4. The FCR values varied between 3.7 and 4.1. All these studies were carried out for Nile tilapia only and the information for other species was not available. The FCR values were very similar even though the studies were carried out in different rearing systems, e.g. cages, earthen ponds and fibre glass tanks. However, considering the highly variable growth responses of Nile tilapia fed hornwort meal, it may not be appropriate to use these FCR values without further verification.

Food conversion ratios for fresh hornwort and oxygen weed fed to grass carp are given in Table 6.6. On a fresh weight basis, the FCR of hornwort varied between 96 and 128, while for oxygen weed it varied between 46 and 132. The apparent variation in FCR values is not surprising, considering the fact that the feeding trials were conducted in different experimental systems and under varying environmental conditions, using fish of different sizes. Devaraj, Maniserry and Keshavappa (1985) reported a fresh weight FCR of 46 for oxygen weed using 3.0 g grass carp in a cement cistern, while Keshavanath and Basavaraju (1980) obtained an FCR value of 132 for oxygen weed in an irrigation canal with 500 g grass carp. Therefore, it is difficult to generalize an FCR value from the available data. Nevertheless, for practical use, the assumption of FCRs of 100-125 for hornwort and 60-100 for oxygen weed on a fresh weight basis may be acceptable.

Digestibility
Digestibility coefficients of hornwort, oxygen weed, *E. canadensis, Najas* spp. and *Ruppia maritima* fed to fish and shrimp (*Penaeus monodon*) are presented in Table

TABLE 6.6
Food conversion ratios of hornwort and oxygen weed fed to grass carp and Nile tilapia

Macrophytes	Fish species	Fish size (g)	Food conversion ratio		References
			Dry weight basis	Fresh weight basis	
Hornwort	Nile tilapia	21.7	15.2	n.s.	Tantikitti *et al.* (1988)
Hornwort	Grass carp	12.0	10.3	128.4	Venkatesh and Shetty (1978a)
Hornwort	Grass carp	14.4	4.1	97.6	Hajra (1987)
Hornwort	Grass carp	52.2	4.05	96.4	Hajra (1987)
Oxygen weed	Grass carp	12.0	9.4	94.0	Venkatesh and Shetty (1978a)
Oxygen weed	Grass carp	3.0	n.s.	45.6	Devaraj, Maniserry and Keshavappa (1985)
Oxygen weed	Grass carp	500.0	n.s.	132.0	Keshavanath and Basavaraju (1980)
Oxygen weed	Grass carp	n.s.	n.s.	62.0	Sutton (1974)

TABLE 6.7
Digestibility of five submerged aquatic macrophytes

Macrophytes	Fish/prawn species	Fish size (g)	Digestibility[1] (%)						References
			DM	CP	EE	NFE	CF	GE	
Hornwort	Grass carp	12.0	n.s.	51.0	69.0	n.s.	39.0	n.s.	Venkatesh and Shetty (1978b)
Hornwort	Grass carp	14.4	51.0	74.9	73.0	51.2	38.2	58.3	Hajra (1987)
Hornwort	Grass carp	52.2	49.0	72.0	69.2	49.0	36.8	55.6	Hajra (1987)
Hornwort	Grass carp	n.s.	49.4	n.s.	n.s.	n.s.	n.s.	n.s.	Lin and Chen (1983, cited by Wee, 1991)
Hornwort	Tilapia rendalli	n.s.	48-59	n.s.	n.s.	n.s.	n.s.	n.s.	Caulton (1982)
Oxygen weed	Grass carp	12.0	n.s.	81.0	82.8	n.s.	43.0	n.s.	Venkatesh and Shetty (1978b)
Oxygen weed	Grass carp	n.s.	67.9	n.s.	n.s.	n.s.	n.s.	n.s.	Lin and Chen (1983, cited by Wee, 1991)
Oxygen weed	Rohu	3.6	n.s.	82.6	42.8	50.0	n.s.	57.1	Ray and Das (1994)
Oxygen weed	Etroplus suratensis	n.s.	41.3	64.3	67.2	n.s.	n.s.	n.s.	De Silva and Perera (1983)
N. guadalupensis	Tilapia zillii	60.2	29.3	75.1	75.9	n.s.	n.s.	45.4	Buddington (1979)
N. graminea	Penaeus monodon	30-40	40-47	n.s.	n.s.	n.s.	n.s.	n.s.	Catacutan (1993)
Ruppia maritima	Penaeus monodon	30-40	70-76	n.s.	n.s.	n.s.	n.s.	n.s.	Catacutan (1993)

[1] DM = dry matter; CP = crude protein; EE = ether extract; CF = crude fibre; NFE = nitrogen free extract; GE = gross energy

6.7. Digestibility coefficients varied between both macrophyte and fish species. Dry matter digestibility appears to be in the range of 40-70 percent, although a rather low value (29 percent) is reported for *N. guadalupensis* when fed to *Tilapia zillii*. Apparent protein digestibility (APD) varied between 64-83 percent with the exception of the 51 percent APD for hornwort reported by Venkatesh and Shetty (1978b) for grass carp.

Crude lipid digestibility coefficients varied between 67 and 83 percent (Table 6.7) with the exception of the 43 percent lipid digestibility of oxygen weed reported for rohu. The digestibility of NFE of hornwort for grass carp was 49-51 percent and that of oxygen weed for rohu was 50 percent. Data on the crude fibre digestibility of hornwort and oxygen weed was available only for grass carp and varied from 37 to 43 percent.

The wide variability in the digestive efficiency of different macrophytes can partly be attributed to the variation in experimental procedures and techniques employed in the studies reviewed. In addition, variation in chemical composition and the physical characteristics of the plants influences digestibility (Buddington, 1979). Nevertheless, for practical purposes, the dry matter, protein, lipid and carbohydrate digestibility may be taken as 40-60 percent, 60-80 percent, 70-80 percent and 50 percent respectively for these common submerged macrophytes.

7. Emergent aquatic macrophytes

Emergent aquatic macrophytes are defined as plants that are rooted in shallow water with vegetative parts emerging above the water surface. It is thought that emergent macrophytes are the most particularly productive of all aquatic macrophytes since they make the best use of all three possible states—with their roots in sediments beneath water and their photosynthetic parts in the air (Westlake, 1963). Westlake (1966) reported the net yield of emergent macrophytes to range from 35 to 85 tonnes DM/ha/year in fertile ponds.

7.1 CLASSIFICATION

There are many different types of emergent macrophytes commonly found in the shallow stagnant waters of tropical and sub-tropical countries of the world. A list of commonly occurring species is presented in Table 7.1. Most of these macrophytes grow naturally; some, however, are used for human consumption and are cultivated.

TABLE 7.1

Common and scientific names of commonly occurring emergent aquatic macrophytes

Scientific name	Family	Common name
Alisma plantago	Alismataceae	Water plantain
Alternanthera philoxeroides	Amaranthaceae	Alligator weed
Cabomba aquatica	Nymphaceae	Aquarium plant
Colocasia chamissonis	Araceae	Swamp taro
Colocasia esculenta	Araceae	Aroids/ Taro
Cyperus esculentus	Cyperaceae	Sedge
Eleocharis dulcis (tuberosa)	Cyperaceae	Sedge/ Chinese water chestnut
Eleocharis ochrostachys	Cyperaceae	Sedge/ Chinese water chestnut
Eleocharis plantagenera	Cyperaceae	Sedge/ Chinese water chestnut
Euryale ferox	Nymphaceae	Water lily
Hydroryza aristata	Gramineae	Swimming grass
Jussiaea repens	Onagraceae	Water primrose
Justicia americana	Acanthaceae	Water willow
Leersia hexandra	Gramineae	Rice cut-grass
Monochoria hastata	Pontederiaceae	Monochoria
Nelumbo nucifera (speciosa)	Nelumbonaceae	Lotus
Nuphar luteum	Nymphaceae	Yellow water lily
Nymphaea lotus	Nymphaceae	Water lily[1]
Nymphaea rubra	Nymphaceae	Red water lily
Nymphaea stellata	Nymphaceae	Blue water lily
Panicum repens	Gramineae	Torpedo grass
Polygonum hydropiper	Polygonaceae	Smart weed
Sagittaria sagittifolia	Alismataceae	Arrowhead
Sagittaria trifolia (sinensis)	Alismataceae	Arrowhead
Scirpus acutus	Cyperaceae	Hardstem bulrush-
Scirpus debilis	Cyperaceae	Weakstalk bulrush
Scirpus mucronatus	Cyperaceae	Ricefield bulrush
Sium sisarum	Apiaceae	Skirret
Sparganium americanum	Sparganiaceae	Bur-reed
Typha latifolia	Typhaceae	Cat tail/ Reed-mace

[1] There are many species of water lily (e.g. *Nymphaea lotus, N. nouchali, N. stellata, Victoria amazonica, V. cruziana*); the most commonly found species is *N. lotus*

7.2 CHARACTERISTICS

Arrowhead (*Sagittaria* spp.) has eight or more underground stems, each with a corm at the end. *S. trifolia* grows wild or semi-cultivated in swamps throughout tropical and subtropical Asia (Ruskin and Shipley, 1976). It is also widely cultivated in China and

FIGURE 7.1
Lotus (*Nelumbo nucifera*), an emergent macrophyte grown in a floodplain, Rangpur, Bangladesh

Lotus leaves are generally quite strong and are often used as disposable plates and as wrapping materials in the countries of south Asia. Please note the support system on the back side (right) of the leave to show its strength

Hong Kong (Herklots, 1972). *S. trifolia* and other species of arrowheads are cultivated by Chinese people in many other parts of the world (Cook *et al.*, 1974). It is reported to be a serious and widespread weed in many countries. However, since it grows quickly and no special care is needed, it could probably be developed into a useful crop. There are no yield data but it can be harvested after 6-7 months (Ruskin and Shipley, 1976). The corms of the arrowhead are boiled like potatoes and are eaten by Chinese and Japanese people with meat dishes. *Sium sisarum* is another emergent macrophyte that is cultivated for its edible roots (Cook *et al.*, 1974). These authors also note that taro (*Colocasia esculenta*) has a starch filled rhizome that is often eaten. Sedge (*Cyperus esculentus*) is widely cultivated for its edible tubers.

The Chinese water chestnut (*Eleocharis dulcis*) is an emergent aquatic plant that grows throughout the year. It is an erect stout and slender perennial leafless sedge (Pandey and Srivastava, 1991a). It has corms or tubers, which are produced in large quantities on underground rhizomes towards the end of the growing season. The plant is widespread from Madagascar to India, SE Asia, Melanesia and Fiji. Occasionally, it is used as a wild source of food in Java and the Philippines. It is cultivated in China for the high starch content of its tubers. It commonly grows in swamps and shallow waters. The yield was said to be greater than 7 tonnes/ha (Ruskin and Shipley, 1976), while Hodge (1956) noted that the yield is much higher, about 18-37 tonnes/ha. Pandey and Srivastava (1991a) reported the promising potential of this plant for leaf protein concentrate (LPC) production.

FIGURE 7.2
Water lily plant covering a substantial part of Lake Awasa, Ethiopia

Courtesy of P.C. Prabu

Monochoria hastata is a robust, fast-growing perennial herb commonly found in ponds, lakes and reservoirs. The fresh biomass yield is about 38-39 tonnes/ha/year (Pandey and Srivastava, 1991b). These authors also reported that this plant also has promising potential for LPC production.

Lotuses and water lilies are common aquatic macrophytes that grow naturally in large natural depressions and lakes, and even in small ditches (Figures 7.1 and 7.2). Lotus flowers have religious significance for Hindus and Buddhists. The lotus is also used for human consumption and is widely cultivated in China and India, mainly for its flowers. The fruits, seeds, rhizomes and stems of water lilies are eaten in S. Asia and India as vegetables and salad.

7.3 PRODUCTION

Although some emergent aquatic macrophytes are consumed by humans and cultivated for this purpose, there is very little in the literature about the way this is practised. However, there are a few reports of experimental observations.

Sutton (1990) cultured *Sagittaria subulata* for 32 weeks in pans filled with sand amended with fertilizers and held in an outdoor tank with flowing pond water. The dry weight of the plants in the highest level of fertilizer (osmocote 35 g/container) was 69 g/m^2. This was 89 percent less than plants collected from a field population (646 ±184 g/m^2) in the Wakulla River, Florida. The water temperature ranged from 16.5 to 46.5 °C during this culture period.

Sharma (1981) reported the growth of *Typha elephantina* in a drain basin (200-300 m wide, 5 km long) of the Agricultural Farm at Jaipur. The area was divided into three zones – submerged, marsh and dry zones depending on the moisture contents. The net annual production in these zones was as follows: dry zone – 1 991 g/m^2/year; marsh zone 2 327 g/m^2/year; and submerged zone 3 696 g/m^2/year.

Camargo and Florentino (2000) studied the seasonal variations in the biomass production of the aquatic macrophyte *Nymphaea rudgeana* in an arm of the Itanhaém River (São Paulo State, Brazil). In November (13.1 g DW/m^2) a gradual increase of biomass was recorded that reached a maximum in February (163.1 g DW/m^2). Then, the biomass decreased, maintaining low levels until a new growth period. The reduction of biomass was associated to the development of floating aquatic macrophytes (*Pistia stratiotes* and *Salvinia molesta*) and, subsequently, to environmental factors (higher salinity values) that were unfavourable to their development. The net primary production of *N. rudgeana* was estimated from the biomass data; the annual productivity value was estimated between 3.02 and 3.82 tonnes/ha/year.

7.4 CHEMICAL COMPOSITION

Analyses of the chemical composition of several emergent macrophytes are presented in Table 7.2. The moisture content of these plants varied between 70 and 92 percent. Generally, the emergent macrophytes have a lower moisture content compared to floating and submerged macrophytes. On a dry matter basis the reported crude protein levels varied between 5 and 40 percent, although most were between 10 and 14 percent. Wide variation is seen in the lipid contents, which ranged between 1.0 and 11 percent. In general, ash contents were relatively low (7 to 20 percent) when compared with other macrophytes. With the exception of taro (*C. esculenta*) and alligator weed (*Alternanthera philoxeroides*) the emergent macrophytes listed in Table 7.2 had high crude fibre levels (20 to 33 percent). Nitrogen free extracts varied between 37 and 53 percent. Amongst these emergent macrophytes, *Sparganium americanum, C. esculenta* and *M. hastata* were particularly rich in protein; being 23.8 percent, 25.0 percent and 39.5 percent (DM), respectively.

TABLE 7.2
Chemical analyses of some common emergent aquatic macrophytes

	Moisture (percent)	Proximate composition[1] (percent DM)					Minerals[1] (percent DM)		Reference
		CP	EE	Ash	CF	NFE	Ca	P	
Alligator weed (*Alternanthera philoxeroides*)	84.1	15.1	2.5	20.1	15.1	47.2	n.s.	n.s.	Tacon (1987)
Aroids/taro (*Colocasia esculenta*), Nigeria	91.8	25.0	10.7	12.4	12.1	39.8	1.74	0.58	Gohl (1981)
Arrow head (*Sagittaria latifolia*), USA	85.0	17.1	6.7	10.3	27.6[3]		0.83[2]	0.35[2]	Boyd (1968)
Bur-reed (*Sparganium americanum*)	89.1	23.8	8.3	11.0	20.2	36.7	n.s.	n.s.	Tacon (1987)
Cat tail (*Typha latifolia*), USA	77.1	10.3	3.9	6.9	33.2[3]		0.64[2]	0.17[2]	Boyd (1968)
Monochoria (*Monochoria hastata*), India	n.s.	39.5	n.s.	7.0	n.s.		n.s.	n.s.	Pandey and Srivasta (1991b)
Rice cut-grass (*Leersia hexandra*), Tanzania	70.0	10.1	1.8	10.4	25.6	52.1	n.s.	n.s.	Gohl (1981)
Sedge/water chestnut (*Eleocharis dulcis*), India	81.1	13.9	n.s.	n.s.	n.s.		n.s.	n.s.	Pandey and Srivasta (1991a)
Sedge/water chestnut (*Eleocharis ochrostachys*), Thailand	n.s.	5.0	1.0	11.9	29.2	52.9	n.s.	n.s.	Klinavee, Tansakul and Promkuntong (1990)
Smart weed (*Polygonum hydropiper*), USA	80.8	11.9	2.4	7.8	26.9[3]		n.s.	n.s.	Boyd (1968)
Torpedo grass (*Panicum repens*), Malaysia	n.s.	14.0	2.1	13.4	32.6	37.9	n.s.	n.s.	Gohl (1981)
Water willow (*Justicia americana*)	85.0	17.6	3.5	16.1	24.0	38.8	0.82	0.12	Tacon (1987)

[1] CP = crude protein, EE = ether extract, CF = crude fibre, NFE = nitrogen free extract, Ca = calcium, P = phosphorus
[2] Data obtained from Tacon (1987)
[3] Cellulose

7.5 USE AS AQUAFEED

Although some of these emergent macrophytes are reasonably rich nutritionally, most are least preferred by macrophytophagous fish. With the exception of one field observation on the utilization of fresh *M. hastata* by grass carp fry in cages, information on their use as fish feed are almost non-existent. Along with some submerged and floating macrophytes, chopped or whole Monochoria are supplied into the cages in the oxbow lakes in southwestern Bangladesh for raising fingerlings. The author of this document has observed that the roots and tender leaves of *M. hastata* are often eaten by grass carp fingerlings.

Similarly, information on the use of dried or processed emergent macrophytes as fish feed is lacking. Venugopal (1980, cited by Shetty and Nandeesha, 1988) found that it was possible to replace fishmeal partly with leaf powder of taro (*C. esculenta*) in feeds for Indian major carp (catla and mrigal) and common carp. In China, alligator weed is used for feeding Chinese carps (Z. Xiaowei, *pers. com.* 2003). Fresh alligator weed is mashed into liquid form with a high-speed beater and applied to the pond for carp fingerlings. Alternatively, mashed alligator weed is mixed with rice bran and fermented overnight before application to the pond. Two per cent table salt is added to eliminate saponin, which is toxic to fish. Alligator weed has also been reported to be cooked and mixed with rice bran before being fed to all the important cultured carps in China (Edwards, 1987). In another study, Klinavee, Tansakul and Promkuntong (1990) used dried sedge/Chinese water chestnut (*Eleocharis ochrostachys*) leaf meal in a pelleted feed for Nile tilapia. These authors incorporated 40 percent dried Chinese water chestnut meal in the diet, in combination with fishmeal (22.25 percent) and rice bran (29.25 percent), which was used to feed the fish in indoor aquaria. Growth performance and food utilization was significantly reduced for fish fed the test diet containing sedge meal when compared with those fed chicken pellets (16.8 percent crude protein).

8. Conclusions

This section provides some overall conclusions from each section of this review.

8.1 ALGAE
From the studies conducted to date it may be concluded that:
- Only about 10-15 percent of dietary protein requirement can be met by algae in test diets without compromising growth and food utilization. There is a progressive decrease in fish performance when dietary incorporation of algal meal rises above 15-20 percent. Total replacement of fishmeal by algal meal generally shows very poor growth responses. Apart from commonly observed impaired growth, the use of algae as the sole source of protein in fish feed can also result in malformation.
- The poor performance of fish fed diets containing higher inclusion levels of algae may be attributable to high levels of carbohydrate, of which only a small fraction consists of mono- and di-saccharides. A preponderance of complex and structural carbohydrates may cause low digestibility.
- The collection, drying and pelletization of algae require considerable time and effort and algal cultivation is costly. Cost-benefit analysis is needed before any definite conclusions on the future application of algae as fish feed can be drawn. The use of algae as fish feed additives may be limited to the commercial production of high value fish.

8.2 *AZOLLA*
The following conclusions can be drawn:
- Laboratory feeding trials on the use of fresh or dried *Azolla* as a complete diet for fish show inconclusive results. Adequate consideration should be given to the preference of each target fish to particular species of *Azolla* before they are used as feed.
- Similarly fresh *Azolla* as a complete diet for high-density cage culture may not be economically viable. However, *Azolla* may be useful in low density and low input cage culture.
- As fish food in *Azolla*-fish pond culture, *Azolla* contributes directly to weight gain of macrophytophagous fish. At the same time, increased production of fish faeces from *Azolla* fodder may be directly consumed by bottom dwellers in addition to being used as an organic (nitrogenous) fertilizer to increase overall pond productivity. However, it should be understood that although the contribution of *Azolla* to aquaculture is interesting, it alone could not ensure high productivity. It can be a useful supplement to natural feed in low-input aquaculture.
- The high rates of decomposition of *Azolla* make it a suitable substrate for enriching the detrital food chain or for microbial processing such as composting, prior to application in ponds.
- The results of several studies indicate that rice-fish-*Azolla* integration increased the yield of both rice and fish compared to rice-fish culture alone. The likely reasons for the increase in rice yield are improved soil fertility resulting from the increased production of fish faeces from *Azolla* fodder; reduced weed growth; and a decreased incidence of insects and pests. Fish yields increase through the direct consumption of *Azolla*.

The advantages of rice-fish-*Azolla* integration may be summarized as follows:
- o increase in fish and rice yields;
- o decrease in need for inorganic fertilizers and pesticides;
- o decrease in incidence of pests and weeds; and
- o improvement of soil fertility.

However, the adoption of rice-fish-*Azolla* integration depends on the attitude and capacity of the farmers, the capacity of support services, including the *Azolla* inoculum availability, and the overall economic feasibility of the system.

8.3 DUCKWEEDS

Duckweeds have received much attention because of their potential to remove mineral contaminants from wastewater. Definitive information has been published on the production and chemical composition of these plants, and their environmental requirements have been clearly determined. Information on the cultivation techniques of many duckweed species is also available. Due to their rapid growth, attractive nutritional properties and relative ease of production, duckweeds have generated renewed interest among fish nutritionists on their use as possible alternative sources of fish feed. It can be concluded that:

- The results of laboratory and field studies and on-farm utilization of these macrophytes clearly indicate that duckweed can provide a complete feed package for carp/tilapia polyculture.
- Successful use of duckweed as fish feed will ultimately depend on the appropriate integration of duckweed production and aquaculture. A preliminary model on duckweed-based aquaculture has been developed and tested under experimental and farming conditions in Bangladesh. However, there is clearly room for fine-tuning this model. Further research towards optimization of the species mix and quantification of feed application for sustainable yield may be necessary.
- The production costs of duckweed, whether as a by-product of wastewater treatment or produced through farming, will ultimately dictate the success of duckweed-based aquaculture. It must be emphasized that sufficient quantities of wastewater may not be available throughout the year to support duckweed production. Therefore, farmers should have the option of using both wastewaters and fertilizers (both chemical and organic) to produce duckweed. The market value of potential fertilizers will eventually determine the economic feasibility of duckweed cultivation. In many countries, including Bangladesh, duckweed cannot be grown all year round because water bodies dry up in the dry season. The availability of fish feed/fertilizer during the dry season needs also to be addressed in a duckweed-based aquaculture model.

8.4 WATER HYACINTHS

A large number of experimental studies have been carried out on the use of fresh or processed water hyacinths as fish feed. In general, water hyacinths have been proven to be moderately successful as a fish feed, although the results are variable. Most of the laboratory studies carried out on the use of water hyacinth as fish feed concentrated on the use of dried meal in pelleted feeds. The results of these studies indicate that water hyacinth leaf meal cannot be used as a fishmeal replacer without compromising growth and food utilization. It has also been noted that, like all other plant ingredients or non-conventional feedstuffs, high dietary inclusion levels (75 percent or above) of water hyacinth meal in complete diets is not feasible, as the minimum dietary protein requirement for most fish species is above 30 percent. Dried water hyacinth leaf meal contains 20-23 percent and whole meal 13-16 percent crude protein (DM). It also must be emphasized that complete diets are not generally used in semi-intensive aquaculture practices in most of the developing countries of the world. Nevertheless, it is clear from the results of many laboratory studies that dried water hyacinth meal has been used successfully as an alternative to rice bran and/or the rice bran-oil cake mixture that is traditionally used as fish feed in many developing countries.

In many of the studies, diets with high water hyacinth inclusion levels performed poorly when compared to fishmeal-based control diets. It must, however, be pointed out that it was only the direct nutritional benefit of water hyacinth that was assessed in these controlled aquarium studies. In natural pond systems, the indirect nutritional value resulting from the production of natural food enhanced by the fertilization of uneaten feed and fish faeces should not be overlooked.

Although few studies report the successful use of fresh water hyacinth as fish feed, it is apparent that the use of processed water hyacinth holds much better promise. However, the question to be answered is: what processing method would be the most viable alternative? If a comparison between water hyacinth processed by different techniques is to be made, we must first consider if the use of dried water hyacinth meal in pelleted diets is feasible under semi-intensive aquaculture in tropical countries. From the experience of the first author of this document, the milling of dried of water hyacinth is labour intensive and pelletization would be even more complicated. Will fish feed manufacturers come forward to use water hyacinth meal to make the feeds cheaper? It is unlikely. Therefore, it is conjectured that the use of dried water hyacinth meal in pelleted feed is not a viable option for tropical small-scale semi-intensive aquaculture. Similarly, the labour-intensive process of drying and milling of water hyacinth may also discourage farmers to use this ingredient in farm-made aquafeed.

On the other hand, water hyacinth processed by composting or fermentation provides similar or higher nutritional benefit but is much less labour intensive. The preparation of water hyacinth paste requires a high-speed blender and the provision of electricity, however, and may therefore be less attractive for smallholder aquaculture. Thus, under the current state of knowledge, it is concluded that composted or fermented water hyacinth used singly or in combination with other traditional dietary ingredients holds promise as a supplemental feed for use in semi-intensive fish culture systems where natural food, produced by fertilization, provides a substantial part of nutrition for fish. The level of its inclusion, when used in combination with other ingredients, will vary and will depend upon its availability, processing costs, the fish species in question and the availability of other ingredients in the locality.

It can therefore be concluded that:
- The use of fresh water hyacinth as an aquafeed is unlikely to be successful.
- The use of dried water hyacinth, though having nutritional value, is unlikely to become viable for use in small-scale aquaculture.
- The use of composted or fermented water hyacinth, however, does hold promise as a dietary ingredient in aquafeeds for small-scale aquaculture.

8.5 OTHER FLOATING MACROPHYTES

Floating macrophytes such as water spinach, water fern, and water lettuce are reasonably rich in protein. However, apart from some general observations (a) in Bangladesh that fresh water spinach is a preferred macrophyte for grass carp and that fresh water lettuce is sometimes given to grass carp and Java barb and (b) that mashed and/ or fermented fresh water lettuce is used in China for feeding carp fingerlings, no detailed investigations have been carried out on any other qualitative aspects of their use.

A limited number of research studies on the use of these other floating macrophytes in pelleted diets for grass carp, common carp and rohu have been carried out. The results generally indicated that higher inclusion rates were not able to produce good growth if the feeds were the only source of nutrition in controlled experimental conditions.

However, when used in a natural or semi-natural rearing system, where plankton and benthos formed a component of the consumption of the fish, diets containing up to 50 percent water lettuce and water fern produced reasonably good growth for grass carp and common carp. These studies also apparently indicate that the FCRs of these macrophytes are reasonably good, varying between 2.0-2.5, when used they are as a dried meal at maximum inclusion level of 50 percent in combination with other proven dietary ingredients/protein sources.

The information currently available is insufficient to draw any definite general conclusions on the suitability of these floating macrophytes for fish feeding in small-scale aquaculture.

8.6 SUBMERGED MACROPHYTES

A large number of research and field studies have been carried out on the utilization of various submerged aquatic macrophytes. These have included monitoring the growth performance, food conversion and digestibility of fish fed different macrophytes; the determination of their nutritional composition; and the consumption and preference of various macrophytes by herbivorous fish. Most of these studies, however, were carried out with grass carp and to a lesser extent with tilapia. The macrophyte species evaluated so far have been oxygen weed, hornwort, pondweeds, *Chara*, water velvet and water milfoil.

Submerged macrophytes were fed to fish either in fresh form or as dried meal in pelleted diets. Reports on other processing techniques designed to improve their nutritional qualities are not available. The number of studies have been conducted to evaluate dried submerged macrophytes as fish feed are extremely limited; so far only dried hornwort meal has been fed to Nile tilapia in a pelleted diet. The results of the above feeding trials were inclusive. Fish reared on fresh submerged macrophytes as the only diet in a clear water rearing system generally produced lower growth responses than fish reared in cement cisterns or ponds where plankton and benthos formed a substantial part of the nutrition of the fish. Submerged aquatic macrophytes usually contain about 13-16 percent crude protein (DM) and were therefore unable to support good growth if used as the only source of dietary protein. However, when submerged macrophytes were fed to fish in natural or semi-natural rearing system, they supported moderate to good growth with fresh weight food conversion ratios varying from 60-125 percent. Most of the fresh submerged macrophytes were well digested by fish, dry matter and protein digestibility being 40-60 percent and 60-80 percent, respectively.

Consumption rates for different macrophytes varied between different fish species. Usually, lower consumption rates were recorded for fish reared in a clear water indoor rearing system. It is also emphasized that the palatability of the same macrophyte species may vary considerably, depending on the environmental conditions under which they grow. Consequently, consumption rates may differ for the same macrophytes for the same fish species owing to variations in environmental conditions, as well as fish size. A wide range of consumption rates of hornwort and oxygen weed were reported for grass carp ranging from 25-200 percent, depending on the fish size. Therefore, to avoid discrepancy, *ad libitum* feeding may preferably be practised for feeding fresh submerged macrophytes.

It is difficult to generalize the preference of submerged macrophytes by different macrophytophagous fish species, because research studies or field observations were conducted under different environmental conditions and in different parts of the world using fish of different age groups. However, it is apparent that grass carp generally prefer fresh soft submerged macrophytes, preferring oxygen weed and *Najas* over *Potamogeton*, *Ceratophyllum*, *Elodea* and *Myriophyllum*. Tender leaves of eelgrass (*Vallisneria*) are also eaten by grass carp. Tilapia (*T. zillii*) feed on *Najas*, *Chara*, *Potamogeton*, *Elodea* and *Myriophyllum*, although it has shown to have preference for *Najas* and *Chara* over other macrophytes. The submerged macrophyte preference of Java barb has not been documented; it generally feeds on *Hydrilla*, *Ceratophyllum* and *Potamogeton*, although it is likely that it will feed other soft fresh macrophytes as well.

8.7 EMERGENT MACROPHYTES

The limited number of observations reported so far is inadequate to draw any conclusions on the use of emergent macrophytes as fish feed; further studies are needed.

9. References

Abdel-Tawwab, M. 2008. The preference of the omnivorous-macrophagous, *Tilapia zillii* (Gervais) to consume a natural free-floating fern, *Azolla pinnata*. *Journal of the World Aquaculture Society*, 39: 104-112.

Alaerts, G.J., Mahbubar, M.R. & Kelderman, P. 1996. Performance of a full-scale duckweed-covered sewage lagoon. *Water Research*, 30: 843-852.

Alalade, O.A. & Iyayi, E.A. 2006. Chemical composition and the feeding value of Azolla (*Azolla pinnata*) meal for egg-type chicks. *International Journal of Poultry Science*, 5: 137-141.

Almazan, G.S., Pullin, R.S.V., Angeles, A.F., Manalo, T.A., Agbayani, R.A. & Trono, M.T.B. 1986. *Azolla pinnata* as a dietary component for Nile tilapia *Oreochromis niloticus*. pp. 523-528. *In* J.L. Maclean, L.B. Dizon & L.V. Hosillos, eds. *The First Asian Fisheries Forum*. Manila, Asian Fisheries Society.

Anonymous. 1980. *Pond fish culture in China*. Lecture notes for FAO training course. Guangzhou, Pearl River Fisheries Research Institute.

Antoine, T., Carraro, S., Micha, J.C. & Van Hove, C. 1986. Comparative appetency for *Azolla* of *Cichlasoma* and *Oreochromis* (*Tilapia*). *Aquaculture*, 53: 95-99.

Appler, H.N. 1985. Evaluation of *Hydrodictyon reticulatum* as protein source in feeds for *Oreochromis* (*Tilapia*) *niloticus* and *Tilapia zillii*. *Journal of Fish Biology*, 27: 327-334.

Appler, H.N. & Jauncey, K. 1983. The utilization of a filamentous green alga (*Cladophora glomerata* (L) Kutzin) as a protein source in pelleted feeds for *Sarotherodon* (*Tilapia*) *niloticus* fingerlings. *Aquaculture*, 30: 21-30.

Aravindakshan, P.K., Jena, J.K., Ayyappan, S., Routray, P., Muduli, H.K., Chandra, S. & Tripathi, S.D. 1999. Evaluation of production trials with grass carp as a major component in carp polyculture. *Journal of the Inland Fisheries Society of India*, 31: 64-68.

Ayyappan, S., Pandey, B.K., Sarkar, S., Saha, D. & Tripathy, S.D. 1991. Potential of *Spirulina* as feed supplement for carp fry. pp. 86-88. *In Proceedings of the National Symposium of Freshwater Aquaculture, CIFA, Bhubaneswar, India*. Bhubaneswar, CIFA.

Azim, M.E., Wahab, M.A., van Dam, A.A., Beveridge, M.C.M. & Verdegem, M.C.J. 2001. The potential of periphyton-based culture of two Indian major carps rohu *Labeo rohita* (Hamilton) and gonia *Labeo gonius* (Linnaeus). *Aquaculture Research*, 32: 209-216.

Azim, M.E., Verdegem, M.C.J., Khatoon, H., Wahab, M.A., van Dam, A.A. & Beveridge, M.C.M. 2002a. A comparison of fertilization, feeding and three periphyton substrates for increasing fish production in freshwater pond aquaculture in Bangladesh. *Aquaculture*, 212: 227-243.

Azim, M.E., Verdegem, M.C.J., Rahman, M.M., Wahab, M.A., van Dam, A.A. & Beveridge, M.C.M. 2002b. Evaluation of polyculture of Indian major carps in periphyton-based ponds. *Aquaculture*, 213: 131-149.

Azim, M.E., Wahab, M.A., Biswas, P.K., Asaeda, T., Fujino, T. & Verdegem, M.C.J. 2004. The effect of periphyton substrate density on production in freshwater polyculture ponds. *Aquaculture*, 212: 441-453.

Bala, N. & Hasan, M.R. 1999. Seasonal fluctuation in water levels and water quality in oxbow lakes in relation to fish yields and social conflict. pp. 163-169. *In* H.A.J. Middendorp, P. Thompson & R.S. Pomeroy, eds. *Sustainable Inland Fisheries Management in Bangladesh*. ICLARM Conference Proceedings No. 58. Manila, ICLARM.

Baur, R.J. & Buck, D.H. 1980. Active research on the use of duckweeds in the culture of grass carp, tilapia, and freshwater prawns. Illinois *Natural History Survey, RR1, Kinmundy*, 111 (*unpublished*).

Becking, J.H. 1979. Environmental requirements of *Azolla* for use in tropical rice production. pp. 345-373. *In Nitrogen and Rice*, Los Banos, Laguna, International Rice Research Institute.

BFRI. 1997. *Research Progress Report: January-August, 1997*. Freshwater Station, Bangladesh Fisheries Research Institute, Mymensingh 2201, Bangladesh (*Unpublished*).

Bhatia, H.L. 1970. Grass carp can control aquatic weeds. *Indian Farming*, 20: 36-37.

Bhaumik, U., Mittal, I.C., Das, P. & Paria, T. 2005. Ecology, periphytic structure and fishery in two floodplain wetlands of West Bengal. *Journal of the Inland Fisheries Society of India*, 37: 54-59.

Bhukaswan, T., Pholprasith, S. & Chatmalai, S. 1981. Aquatic weed control by the grass carp. *Thai Fisheries Gazette*, 34: 529-538.

Bolenz, S., Omran, H. & Gierschner, K. 1990. Treatments of water hyacinth tissue to obtain useful products. *Biological Wastes*, 33: 263-274.

Bonar, S.A., Sehgal, H.S., Pauley, G.B. & Thomas, G.L. 1990. Relationship between the chemical composition of aquatic macrophytes and their consumption by grass carp, *Ctenopharyngodon idella*. *Journal of Fish Biology*, 36: 149-157.

Borlongan, I.G. & Coloso, R.M. 1994. Leaf meals as protein sources in diets for milkfish *Chanos chanos* (Forsskal). pp. 63-68. *In* S.S. De Silva, ed. *Fish Nutrition Research in Asia, Special Publication No. 6*. Manila, Asian Fisheries Society.

Boyd, C.E. 1968. Fresh-water plants: a potential source of protein. *Economic Botany*. 22: 359-368.

Briggs, M.R.P. & Funge-Smith, S.J. 1996. The potential use of *Gracilaria* sp. meal in diets for juvenile *Penaeus monodon* Fabricius. *Aquaculture Research*, 27: 345-354.

Buckingham, K.W., Ela, S.W., Morris, J.G. & Goldman, C.R. 1978. Nutritive value of the nitrogen-fixing aquatic fern *Azolla filiculoides*. *Journal of Agricultural and Food Chemistry*, 26: 1230-1234.

Buddington, R.K. 1979. Digestion of an aquatic macrophyte by *Tilapia zillii* (Gervais). *Journal of Fish Biology*, 15: 449-455.

Buddington, R.K. 1980. Hydrolysis-resistant organic matter as a reference for measurement of fish digestion efficiency. *Transactions of the American Fisheries Society*, 109: 653-656.

Cagauan, A.G. 1994. Azolla in rice-fish farming system. pp. 42-45. *In* L.M. Chou, A.D. Munro, T.J. Lam, T.W. Chen, L.K.K. Cheong, J.K. Ding, K.K. Hooi, H.W. Khoo, V.P.E. Phang, K.F. Shim & C.H. Tan, eds. *The Third Asian Fisheries Forum*, Manila, Asian Fisheries Society.

Cagauan, A.G. & Nerona, V.C. 1986. Tilapia integrated rice-fish culture with *Azolla* as biofertilizer. *Fisheries Research Journal of the Philippines*, 11: 29-33.

Cagauan, A.G. & Pullin, R.S.V. 1991. Azolla in aquaculture: Past, present and future. pp. 104-130. *In* J. Muir & R.J. Roberts, eds. *Recent Advances in Aquaculture*. Oxford, Blackwell Science.

Camargo, A.F.M. & Florentino, E.R. 2000. Population dynamics and net primary production of the aquatic macrophyte *Nymphaea rudgeana* C. F. Mey in a lotic environment of the Itanhaem River basin (SP, Brazil). *Revista Brasileira de Biologica*, 60: 83-92.

Cassani, J.R. 1981. Feeding behaviour of underyearling hybrids of the grass carp, *Ctenopharyngodon idella* (female) and the bighead carp, *Hypopthalmichthys nobilis* (male), on selected species of aquatic plants. *Journal of Fish Biology*, 18: 127-133.

Cassani, J.R. & Caton, W.E. 1983. Feeding behaviour of yearling and older hybrid grass carp. *Journal of Fish Biology*, 22: 35-41.

Cassani, J.R., Caton, W.E. & Hansen, T.H. 1982. Culture and diet of hybrid carp fingerlings. *Journal of Aquatic Plant Management*, 20: 30-32.

Catacutan, M.R. 1993. Assimilation of aquatic macrophytes in *Penaeus monodon. Journal of Aquaculture in the Tropics*, 8: 9-12.

Caulton, M.S. 1982. Feeding, metabolism and growth of tilapias: some quantitative considerations. pp. 157-180. *In* R.S.V. Pullin & R.H. Lowe-McConnell, eds. *The Biology and Culture of Tilapias, ICLARM Conference Proceedings* No. 7. Manila, International Centre for Living Aquatic Resources Management.

Chiayvareesajja, S. & Tansakul, R. 1989. Culture of banana prawn (*Penaeus merguiensis*) and tilapia (*Oreochromis niloticus*) by using aquatic weed mixture pellet. pp. 153-156. *In* S.S. De Silva, ed. *Fish Nutrition Research in Asia*. Special Publication No. 4. Manila, Asian Fisheries Society.

Chiayvareesajja, S., Wongwit, C. & Tansakul, R. 1990. Cage culture of tilapia (*Oreochromis niloticus*) using aquatic weed-based pellets. pp. 287-290. *In* R. Hirano & I. Hanyu, eds. *The Second Asian Fisheries Forum*. Manila, Asian Fisheries Society.

Chiayvareesajja, S., Sirikul, B., Sirimontrapon, P., Rakkeaw, S. & Tansakul, R. 1988. Comparison between natural feeding alone and supplemental feeding with pellets containing locally available ingredients for cage culture of *Oreochromis niloticus* in Thale Noi, Thailand. pp. 323-327. *In* R.S.V. Pullin, T. Bhukaswan, K. Tonguthai & J.L. Maclean, eds. *The Second International Symposium on Tilapia in Aquaculture*, ICLARM Conference Proceedings No. 15. Manila, Philippines,

Chiayvareesajja, S., Wongwit, C., Cronin, A., Supamataya, K., Tantikitti, C. & Tansakul, R. 1989. Utilization of aquatic weed mixture pellets as feed for Nile tilapia (*Oreochromis niloticus*) and pig. pp. 143-147. *In* S.S. De Silva, ed. *Fish Nutrition Research in Asia*, Special Publication No. 4. Manila, Asian Fisheries Society.

Chow, C.Y. & Woo, N.Y.S. 1990. Bioenergetic studies on an omnivorous fish, *Oreochromis mossambicus*: evaluation of utilization of *Spirulina* algae in feed. pp. 291-294. In R. Hirano & I. Hanyu, eds. *The Second Asian Fisheries Forum*. Manila, Asian Fisheries Society.

CIFA. 1981. *Annual Report of the Central Inland Fisheries Research Institute*, Barrackpore.

Coche, A.G. 1983. *Freshwater aquaculture development in China. Report of the FAO/UNDP study tour organized for French-speaking African countries, 22 April - 20 May 1980.* FAO Fisheries Technical Paper No. 215. Rome, FAO.

Cole, D.J.A. & Van Lunen, T.A. 1994. Ideal amino acid patterns. pp. 99-112. *In* J.P.I. D'Mello, ed., *Amino acids in farm animal nutrition*. Edinburgh, The Scottish Agricultural College.

Cook, C.D.K., Gut, B.J., Rix, E.M., Schneller, J. & Seitz, M. 1974. *Water Plants of the World. A Manual for the Identification of the Genera of Freshwater Macrophytes*. The Hague, Junk. 561 pp.

Crawford, D.J., Landolt, E., Les, D.H., Archibald, J.K. & Kimball, R.T. 2005. Alloenzyme variation within and divergence between *Lemna gibba* and *L. disperma*: systematic and biogeographic implications. *Aquatic Botany*, 83: 119-128.

Culley, D.D. & Epps, A.E. 1973. Use of duckweeds for waste treatment and animal feed. *Journal of the Water Pollution Control Federation*, 45: 337-347.

Culley, D.D. & Myers, R.W. 1980. *Effect of harvest rate on duckweed yield and nutrient extraction in dairy waste lagoons*. US Department of Energy Final Report. Baton Rouge, School of Forestry and Wildlife Management, Louisiana State University. 6 pp.

Culley, D.D., Gholson, J.H., Chisholm, T.S., Standifer, L.C. & Epps, E.A. 1978. *Water quality renovation of animal waste lagoons utilizing aquatic plants*. EPA Publication No. 600/2-78-153. Ada, Oklahoma, US Environmental Protection Agency. 153 pp.

Culley, D.D., Réjmenková, E., Kvet, J. & Frye, J.B. 1981. Production, chemical quality and use of duckweeds (*Lemnaceae*) in aquaculture, waste management, and animal feeds. *Journal of the World Mariculture Society*, 12(2): 27-49.

Davies, S.J., Brown, M.T. & Camilleri, M. 1997. Preliminary assessment of the seaweed *Porphyra purpurea* in artificial diets for thick-lipped grey mullet (*Chelon labrosus*). *Aquaculture*, 152: 249-258.

Dempster, P.W., Beveridge, M.C.M. & Baird, D.J. 1993. Herbivory in tilapia *Oreochromis niloticus* (L.): a comparison of feeding rates on periphyton and phytoplankton. *Journal of Fish Biology*, 43: 385-392.

De Silva, S.S. 1995. Supplementary feeding in semi-intensive aquaculture systems. pp. 24-60. *In* M.B. New, A.G.J. Tacon & I. Csavas, eds. *Farm-made Aquafeeds*. FAO Fisheries Technical Paper No. 343. Rome, FAO.

De Silva, S.S. & Perera, M.K. 1983. Digestibility of an aquatic macrophyte by the cichlid *Etroplus suratensis* (Bloch) with observations on the relative merits of three indigenous components as markers and daily changes in protein digestibility. *Journal of Fish Biology*, 23: 675-684.

Devaraj, K.V., Krishna, D.V. & Keshavappa, G.Y. 1981. Utilisation of duckweed and waste cabbage leaves in the formulation of fish feed. *Mysore Journal of Agricultural Sciences*, 15: 132–135.

Devaraj, K.V., Manissery, J.K. & Keshavappa, G.K. 1985. On the growth of grass carp (*Ctenopharyngodon idella*) fed with lucerne (*Medicago sativa*) and hydrilla (*Hydrilla verticillata*) *ad libitum* in cement cistern. *Mysore Journal of Agricultural Science*, 19: 275-278.

Dey, S.C. & Sarmah, S. 1982. Prospect of the water hyacinth (*Eichhornia crassipes*) as feed to cultivable fishes - a preliminary study with *Tilapia mossambica* Peters. *Matsya*, 8: 40-44.

Duthu, G.S. & Kilgen, R.H. 1975. Aquarium studies on the selectivity of 16 aquatic plants as food by fingerling hybrids of the cross between *Ctenopharyngodon idella* and *Cyprinus carpio*. *Journal of Fish Biology*, 7: 203-208.

DWRP, 1997. *Literature Review 1*. Dhaka, Duckweed Research Project, Ministry of Fisheries and Livestock and DHN Consultants. 68 pp.

DWRP, 1998. *Duckweed in Bangladesh*. Dhaka, Duckweed Research Project, Ministry of Fisheries and Livestock and Royal Netherlands Embassy. 91 pp.

Edwards, D.J. 1974. Weed preference and growth of young grass carp in New Zealand. *New Zealand Journal of Marine and Freshwater Research*, 8: 341-350.

Edwards, P. 1980. *Food Potential of Aquatic Macrophytes*. Manila, ICLARM Studies and Reviews No. 5. 51 pp.

Edwards, P. 1987. Use of terrestrial vegetation and aquatic macrophytes in aquaculture. pp. 311-335. *In* D.J.W. Moriarty & R.S.V. Pullin, eds. *Detritus and Microbial Ecology in Aquaculture*. Manila, ICLARM Conference Proceedings No. 14.

Edwards, P. 1990. An alternative excreta-reuse strategy for aquaculture: the production of high-protein animal feed. pp. 209-221. *In* P. Edwards & R.S.V. Pullin, eds. *Wastewater-fed Aquaculture*, Bangkok, Environmental Sanitation Information Center, Asian Institute of Technology.

Edwards, P., Kamal, M. & Wee, K.L. 1985. Incorporation of composted and dried water hyacinth in pelleted feed for the tilapia *Oreochromis niloticus* (Peters). *Aquaculture and Fisheries Management*, 16: 233-248.

Edwards, P., Pacharaprakiti, C. & Yomjinda, M. 1990. Direct and indirect use of septage for culture of Nile tilapia *Oreochromis niloticus*. pp. 165-168. *In* R. Hirano & I. Hanyu, eds. *The Second Asian Fisheries Forum*, Manila, Asian Fisheries Society.

Edwards, P., Hassan, M.S., Chao, C.H. & Pacharaprakiti, C. 1992. Cultivation of duckweeds in septage-loaded earthen ponds. *Bioresource Technology*, 40: 109-117.

Effiong, B.N., Sanni, B.N. & Sogbesan, O.A. 2009. Comparative studies on the binding potential and water stability of duckweed meal, corn starch and cassava starch. *New York Science Journal*, 2: 50-57.

El-Sayed, A.F.M. 1992. Effects of substituting fish meal with *Azolla pinnata* in practical diets for fingerling and adult Nile tilapia, *Oreochromis niloticus* L. *Aquaculture and Fisheries Management*, 23: 167-173.

El-Sayed, A.F.M. 2003. Effects of fermentation methods on the nutritive value of water hyacinth for Nile tilapia *Oreochromis niloticus* (L.) fingerlings. *Aquaculture*, 218: 471-478.

El-Sayed, A.F.M. 2008. Effects of substituting fish meal with *Azolla pinnata* in practical diets for fingerling and adult Nile tilapia, *Oreochromis niloticus* (L.). *Aquaculture Research*, 23: 167-173.

Fagbenro, O.A. 1990. Food composition and digestive enzymes in the gut of pond cultured *Clarias isheriensis*. *Journal of Applied Ichthyology*, 6: 91-98.

Fagbenro, O.A., Akinbulumo, M.A. & Ojo, S.O. 2004. Aquaculture in Nigeria – history, status and prospects. *World Aquaculture*, 35(2): 20-23.

FAO. 1977. China: recycling of organic wastes in agriculture. *FAO Soils Bulletin*, 40: 29-40.

FAO. 2008. Report of the FAO Expert Workshop on the Use of Wild Fish and/or Other Aquatic Species as Feed in Aquaculture and its Implications to Food Security and Poverty Alleviation, Kochi, India, 16-18 November 2007. FAO Fisheries Report No. 867. Rome, FAO.

FAO. 2009. Fisheries Department, Fishery Information, Data and Statistics Unit. Fishstat Plus: Universal software for fishery statistical time series. Version 2006-11-02. Release data 02 March 2006 (available at: www.fao.org/fi/statist/FISOFT/FISHPLUS.asp).

Farhangi, M. & Carter, C.G. 2001. Growth, physiological and immunological responses of rainbow trout (*Oncorhynchus mykiss*) to different dietary inclusion levels of dehulled lupin (*Lupinus angustifolius*). *Aquaculture Research*, 32: 329-340.

Fasakin, E.A., Balogum, A.M. & Fasuru, B.E. 1999. Use of duckweed, *Spirodela polyrrhiza*, L. Schleiden, as a protein feedstuff in practical diets for tilapia, *Oreochromis niloticus* L. *Aquaculture Research*, 30: 313-318.

Ferdoushi, Z., Haque, F., Khan, S. & Haque, M. 2008. The effects of two aquatic floating macrophytes (*Lemna* and *Azolla*) as biofilters of nitrogen and phosphate in fish ponds. *Turkish Journal of Fisheries and Aquatic Sciences*, 8: 253-258.

Ferentinos, L., Smith, J. & Valenzuela, H. 2002. Azolla. Mānoa, College of Tropical Agriculture and Human Resources, University of Hawai'i at Mānoa.

Fiogbé, E.D., Micha, J.C. & Van Hove, C. 2004. Use of a natural aquatic fern, *Azolla microphylla*, as a main component in food for the omnivorous-phytoplanktonophagous tilapia, *Oreochromis niloticus* L. *Journal of Applied Ichthyology*, 20: 517-520.

Fischer, Z. 1968. Food selection in grass carp *Ctenopharyngodon idella* under experimental conditions. *Polskie Archiwum Hydrobiologii*, 15: 1-8.

Fischer, Z. 1970. The elements of energy balance in grass carp (*Ctenopharyngodon idella*). Part I. *Polskie Archiwum Hydrobiologii*, 17: 421-434.

Franceschi, V.R. 1989. Calcium oxalate formation is a rapid and reversible process in *Lemna minor* L. *Protoplasma*, 148: 130-137.

Freidel, J.W. & Bashir, M.O. 1979. On the dynamics of populations and distribution of water hyacinth in the White Nile, Sudan. pp. 94-105. *In* W. Koch (editor), *Weed Research in Sudan*. Sudan, University of Gezira, Wad Medani.

Gaigher, I.G. & Short, R. 1986. An evaluation of duckweed (*Lemnaceae*) as a candidate for aquaculture in South Africa. pp. 81-90. *In* R.D. Wamsley and J.G. Wan, eds. Canberra, CSIRO, Aquaculture 1980 Report Series No. 15.

Gaigher, I.G., Porath, D. & Granoth, G. 1984. Evaluation of duckweed (*Lemna gibba*) as feed for tilapia (*Oreochromis niloticus* X *aureus*) in a recirculation unit. *Aquaculture*, 41: 235-244.

Galkina, N.V. Abdullaev, D.A. & Zacharova, V.L. 1965. Biological and feed features of duckweeds. *Uzbekistan Biological Journal*, 3: 44-47.

Garg, S.K. 2005. Role of periphyton in development of sustainable aquaculture technology for inland saline groundwater: a review. *Indian Journal of Animal Science*, 75: 1348-1353.

Gavina, L.D. 1994. Pig-duck-fish-azolla integration in La Union, Philippines. NAGA, *The ICLARM Quarterly*, 17(2): 18-20.

Gohl, B. 1981. *Tropical Feeds*. Rome, FAO. 529 pp.

Gopal, B. & Chamanlal, 1991. Distribution of aquatic macrophytes in polluted water bodies and their bio-indicator value. *Verhandlungen des Internationalen Vereins für Limnologie*, 24: 2125-2129.

Guha, R. 1997. *Duckweeds. Envis Newsletter*, March 1997: 5-9. Bangalore, Indian Institute of Science.

Gunnarsson, C.C. & Peterson, C.M. 2007. Water hyacinths as a resource in agriculture and energy production: a literature review. *Waste Management*, 27: 117-129.

Habib, M.A.B., Parvin, M., Huntington, T.C. & Hasan, M.R. 2008. A review on culture, production and use of spirulina as food for humans and feeds for domestic animals and fish. *FAO Fisheries and Aquaculture Circular* No. 1034. FAO, Rome. 33 pp.

Hadiuzzaman, S. & Khondker, M. 1993. *Salvinia auriculata* Aublet - a new record of aquatic Pteridophyte from Bangladesh. *Bangladesh Journal of Botany*, 22: 229-231.

Hajra, A. 1987. Biochemical investigations on the protein-calorie availability in grass carp (*Ctenopharyngodon idella* Val.) from an aquatic weed (*Ceratophyllum demersum* Linn.) in the tropics. *Aquaculture*, 61: 113-120.

Hajra, A. & Tripathi, S.D. 1985. Nutritive value of aquatic weed, *Spirodela polyrrhiza* (Linn.) in grass carp. *Indian Journal of Animal Science*, 55: 702-705.

Hall, J. & Payne, G. 1997. Factors controlling the growth of field population of *Hydrodictyon reticulatum* in New Zealand. *Journal of Applied Phycology*, 9: 229-236.

Haller, W.T., Sutton, D.L. & Burlowe, W.C. 1974. Effects of salinity on the growth of several aquatic macrophytes. *Ecology*, 55: 891-894.

Harvey, R.M. & Fox, J.L. 1973. Nutrient removal using *Lemna minor*. *Journal of the Water Pollution Control Federation*, 45: 1928-1938.

Hasan, M.R. 1990. Evaluation of leucaena and water hyacinth leaf meal as dietary protein sources for the fry of Indian major carp, *Labeo rohita* (Hamilton). pp. 209-221. *In* M.H. Mian, ed. *Proceedings of BAU Research Progress* 4, Mymensingh, Bangladesh Agricultural University.

Hasan, M.R. & Middendorp, H.A.J. 1998. Optimizing stocking density of carp fingerlings through modelling of the carp yield in relation to average water transparency in enhanced fisheries in semi-closed water in western Bangladesh. pp. 159-168. *In* T. Petr, ed. *Inland Fishery Enhancements*. FAO Fisheries Technical Paper No. 374, Rome, FAO.

Hasan, M.R. & Roy, P.K. 1994. Evaluation of water hyacinth leaf meal as dietary protein source for Indian major carp, *Labeo rohita* fingerlings. pp. 671-674. *In* L.M. Chou, A.D. Munro, T.J. Lam, T.W. Chen, L.K.K. Cheong, J.K. Ding, K.K. Hooi, H.W. Khoo, V.P.E. Phang, K.F. Shim & C.H. Tan, eds. *The Third Asian Fisheries Forum*. Manila, Asian Fisheries Society.

Hasan, M.R., Moniruzzaman, M. & Omar Farooque, A.M. 1990. Evaluation of leucaena and water hyacinth leaf meal as dietary protein sources for the fry of Indian major carp, *Labeo rohita* (Hamilton). pp. 275-278. *In* R. Hirano & I. Hanyu, eds. *The Second Asian Fisheries Forum*. Manila, Asian Fisheries Society.

Hassan, M.S. & Edwards, P. 1992. Evaluation of duckweed (*Lemna perpusila* and *Spirodela polyrrhiza*) as feed for Nile tilapia (*Oreochromis niloticus*). *Aquaculture*, 104: 315-326.

Haustein, A.T., Gilman, R.H., Skillicorn, P.W., Vergara, V., Guevara, V. & Gastanaduy, A. 1988. Duckweed, a useful strategy for feeding chickens: performance of layers fed with sewage-grown Lemnaceae species. *Poultry Science*, 69: 1835-1844.

Hem, S. & Avit, J.L.B. 1994. First results on 'acadjas enclose' as an extensive aquaculture system (West Africa). *Bulletin of Marine Science*, 55: 1038-1049.

Hensen, R.H. 1990. Spirulina algae improves Japanese fish feeds. *Aquaculture Magazine*, 6(6); 38-43.

Hepher, B. & Pruginin, Y. 1979. *Guide to Fish Culture in Israel: 4. Fertilization, manuring and feeding*. Israel, Foreign Training Department, Israel. 61 pp.

Herfjord, T., Osthagen, H. & Saelthun, N.R. 1994. *The Water Hyacinth*. Oslo, Norwegian Agency for Development Cooperation. 39 pp.

Herklots, G.A.C. 1972. *Vegetables in South-east Asia*. London, George Allen & Unwin. 525 pp.

Hertrampf, J.W. & Piedad-Pascual, F. 2000. Handbook on ingredients for aquaculture feeds. Dordrecht, Kluwer Academic Publishers. 573 pp.

Hickling, D.F. 1966. On the feeding process in the white amur, *Ctenopharyngodon idella*. *Journal of Zoology*, 148: 408-419.

Hillman, W.S. & Culley, D.D. 1978. The uses of duckweed. *American Scientist*, 66: 442-451.

Hodge, W. 1956. Chinese water chestnut or Matai - a paddy crop of China. *Economic Botany*, 10: 49-65.

Huntington, T.C. and Hasan, M.R. 2009. Fish as feed inputs for aquaculture - practices, sustainability and implications: a global synthesis. pp 1-61. *In* M.R. Hasan and M. Halwart, eds. Fish as feed inputs for aquaculture: practices, sustainability and implications. *FAO Fisheries and Aquaculture Technical Paper*. No. 518. Rome, FAO.

Hutabarat, J., Syarani, L. & Smith, A.K.M. 1986. The use of freshwater hyacinth *Eichhornia crassipes* in cage culture in Lake Rawa Penang, Central Java. pp. 570-580. *In* J.L. Maclean, L.B. Dizon & L.V. Hosillos, eds. *The First Asian Fisheries Forum*. Manila, Asian Fisheries Society.

Imaoka, T. & Teranishi, S. 1988. Rates of nutrient uptake and growth of the water hyacinth (*Eichhornia crassipes* (Mart. Solms). *Water Research*, 22: 943-951.

Islam, A.B.M.S. & Haque, M.Z. 1986. Growth of Azolla in association with rice crop culture and its contribution to soil fertility. *Bangladesh Journal of Agriculture*, 11: 87-90.

Islam, A.K.M.N. & Khondker, M. 1991. Preliminary limnological investigations of some polluted waters covered by duckweeds. *Bangladesh Journal of Botany*, 20: 73-75.

Islam, A.K.M.N. & Paul, S.N. 1977. Limnological studies on *Wolffia arrhiza* (L) Wimm. *Journal of the Asiatic Society Bangladesh (Sciences)*, 3: 111-123.

Jagdish, M., Rana, S.V.S. & Agarwal, V.P. 1995. Efficacy of grass carp (*Ctenopharyngodon idella*) in weed control and its growth in Karna Lake (Haryana). *Journal of the Inland Fisheries Society of India*, 27: 49-55.

Jana, S.N., Garg, S.D., Thirunavukkarasu, A.R., Bhatnagar, A., Kalla A. & Patra, B.C. 2006. Use of additional substrate to enhance growth performance of milkfish, *Chanos chanos* (Forsskal) in inland saline groundwater ponds. *Journal of Applied Aquaculture*, 18: 1-20.

Jones, I.D. 1975. Effect of processing by fermentation on nutrients. pp. 324-254. *In* R.S. Harris & K. Karmas, eds. *Nutritional Evaluation of Food Processing*. Westport Connecticut, Avi Publishing.

Juliano, R.O. 1985. The biology of milk fish (*Chanos chanos*, Forsskal) and ecology and dynamics of brackishwater ponds in Philippines. University of Tokyo, Japan. (Dissertation)

Kalla, A., Yoshimatsu, T., Araki, T., Zhang, D., Yamamoto, T. & Sakamoto, S. 2008. Use of *Porphyra* spheroplasts as feed additive for red sea bream. *Fisheries Science*, 74: 104-108.

Kaul, V. & Bakaya, U. 1973. The noxious, floating, lemnid *Salvinia* aquatic weed complex in Kashmir. pp. 183-192. *In* C.R. Varshney & J. Rzòska, eds. *Aquatic Weeds in S.E. Asia*. The Hague, Junk.

Keshavanath, P. & Basavaraju, Y. 1980. A note on the utility of grass carp, *Ctenopharyngodon idella* (Valenciennes) in controlling the aquatic weed, *Hydrilla*. *Current Research*, 9: 154-156.

Keshavanath, P., Gangadhar, B., Ramesh, T.J., van Rooij, J.M., Beveridge, M.C.M., Baird, D.J., Verdegem, M.C.J. & van Dam, A.A. 2001. Use of artificial substrates to enhance production of freshwater herbivorous fish in pond culture. *Aquaculture Research*, 32: 189-197.

Keshavanath, P., Gangadhar, B., Ramesh, T.J., van Dam, A.A., Beveridge, M.C.M. & Verdegem, M.C.J. 2002. The effect of periphyton and supplemental feeding on the production of indigenous carps *Tor khudree* and *Labeo fimbriatus*. *Aquaculture*, 119: 175-190.

Khan, M.A.H. & Haque, M.S. 1991. Factors affecting the growth of *Azolla* - a review. *Bangladesh Journal of Aquaculture*, 11-13: 5-8.

Khondker, M., Islam, A.K.M.N. & Makhnun, A.D. 1994. *Lemna perpusilla*: screening on habitat limnology. *Bangladesh Journal of Botany*, 23: 99-106.

Khondker, M., Islam, A.K.M.N. & Nahar, N. 1993a. Study on the biomass of *Spirodela polyrrhiza* and the related limnological factors of some polluted waters. pp. 37-40. *In* M.S. Khan, M.A. Aziz Khan, S. Hadiuzzaman & A. Aziz, eds. *Plants for the Environment, Proceedings of the 7th Botanical Conference, 13-14 December, 1992*. Dhaka, Bangladesh Botanical Society, Dhaka, Bangladesh.

Khondker, M., Islam, A.K.M.N. & Nahar, N. 1993b. A preliminary study on the growth rate of *Spirodela polyrrhiza*. *Dhaka University Journal of Biological Sciences*, 2: 197-200.

Kim, K-W, Bai, S.C., Koo, J-W. & Wang, X. 2002. Effects of dietary *Chlorella ellipsoidea* supplementation on growth, blood characteristics, and whole-body composition in juvenile Japanese flounder *Paralichthys olivaceus*. *Journal of the World Aquaculture Society*, 33: 425-431.

Klinavee, S., Tansakul, R. & Promkuntong, W. 1990. Growth of Nile tilapia (*Oreochromis niloticus*) fed with aquatic plant mixtures. pp. 283-286. *In* R. Hirano & I. Hanyu, eds. *The Second Asian Fisheries Forum*. Manila, Asian Fisheries Society.

Knipling, E.B., West, S.H. & Haller, W.T. 1970. Growth characteristics, yield potential and nutritive content of water hyacinth. *Proceedings of the Soil and Crop Science Society of Florida*, 30: 51-63.

Kola, K. 1988. Aspects of the ecology of water hyacinth *Eichhornia crassipes* (Martius) Solms. in the Lagos Lagoon System. pp. 80-84. *In* T.A. Farri, ed. *Proceedings of the International Workshop on Water Hyacinth - Menace and Resource*. Lagos, Nigerian Federal Ministry of Science and Technology.

Konan-Brou, A.A. & Guiral, D. 1994. Available algal biomass in tropical brackish water artificial habitats. *Aquaculture*, 119: 175-190.

Konyeme, J.E., Sogbesan, A.O. & Ugwumba, A.A.A. 2006. Nutritive value and utilization of water hyacinth (*Eichhornia crassipes*) meal as plant protein supplement in the diet of *Clarias gariepinus* (Burchell, 1822) (Pisces: Clariidae) fingerlings. *African Scientist*, 7: 127-133.

Kostman, T.A., Tarlyn, N.M., Loewus, F.A. & Franceschi, V.R. 2001. Biosynthesis of L-ascorbic acid and conversion of carbon 1 and 2 of L-ascorbic acid to oxalic acid occurs within individual calcium oxalate crystal idioblasts. *Plant Physiology*, 125: 634-640.

Kumar, K., Ayyappan, S., Murjani, G. & Bhandari, S. 1991. Utilization of mashed water hyacinth as feed in carp rearing. pp. 89-91. *In Proceedings of the National Symposium on Freshwater Aquaculture*, CIFA Bhubaneswar, India. Bhubaneswar, CIFA.

Lahser, C.W. 1967. *Tilapia mossambica* as a fish aquatic weed control. *Progressive Fish Culturist*, 29: 48-50.

Landolt, E. 1986. *The family of* Lemnaceae - *a monographic study: morphology, karyology, ecology, geographic distribution, systematic position, nomenclature, descriptions, Vol. 2.* Zurich, Veroffentlichungen des Geobotanisches Institut der Edg. Tech. Hochschule, Stiftung Ruebel. 566 pp.

Landolt, E. 2006. Duckweed. *In* Flora of North America Editorial Committee, eds. *Flora of North America North of Mexico Vol. 22.* New York and Oxford. 143 pp.

Landolt, E. & Kandeler, R. 1987. *The family of* Lemnaceae - *a monographic study: phytochemistry, physiology, application and bibliography, Vol. 4.* Zurich, Veroffentlichungen des Geobotanisches Institut der Edg. Tech. Hochschule, Stiftung Ruebel. 638 pp.

Lapointe, B.E. & O'Connell, J. 1989. Nutrient-Enhanced growth of *Cladophora prolifera* in Harrington Sound, Bermuda: eutrophication of a confined, phosphorus-limited marine ecosystem. *Estuarine Coastal and Shelf Science ECSSD3*, 28: 347-360.

Lavens, P. & Sorgeloos, P. 1996. Manual on the production and use of live food for aquaculture. FAO Fisheries Technical Paper No. 361. Rome, FAO.

Leng, R.A., Stambolie, J.H. & Bell, R. 1995. Duckweed - a potential high-protein feed resource for domestic animals and fish. pp. 103-114. *In* Proceedings of the 7th Animal Science Congress of the Asian-Australasian Association of Animal Production Societies (AAAP) Conference, Bali. Jakarta, Indonesian Society of Animal Science.

Liang, J.K. and Lovell, R.T. 1971. Nutritional value of water hyacinth in channel catfish feeds. *Hyacinth Control Journal*, 9: 40-44.

Liao, W.L., Takeuchi, T., Watanabe, T. & Yamaguchi, K. 1990. Effect of dietary *Spirulina* supplementation on extractive nitrogenous constituents and sensory test of cultured striped jack flesh. *Journal of the Tokyo University of Fisheries*, 77: 241-246.

Little, D. & Muir, J. 1987. *A Guide to Integrated Warm Water Aquaculture.* Stirling, Institute of Aquaculture Publications, University of Stirling. 238 pp.

Liu, X., Min, C., Xia-shi, L. & Chungchu, L. 2008. Research on some functions of *Azolla* in CELSS system. *Acta Astronautica*, 63: 1061-1066.

Lüönd, A. 1980. Effects of nitrogen and phosphorus upon the growth of some *Lemnaceae*. pp 118-141. *In* E. Landolt, ed. *Biosystematic Investigations in the Family of Duckweeds (Lemnaceae).* Zurich, Veroffentlichungen des Geobotanisches Institut der Edg. Tech. Hochschule, Stiftung Ruebel.

McHugh, D.J. 2002. *Prospects for seaweed production in developing countries.* FAO Fisheries Circular No. 968. Rome, FAO. 28 pp.

McHugh, D.J. 2003. *A guide to the seaweed industry.* FAO Fisheries Technical Paper No. 441. Rome, FAO. 118 pp.

Majhi, S.K., Das, A. & Mandal, B.K. 2006. Growth performance and production of organically cultured grass carp *Ctenopharyngodon idella* (Val.) under mid-hill conditions of Meghalaya; North Eastern India. *Turkish Journal of Fisheries and Aquatic Sciences*, 6: 105-108.

Majid, F.Z., Khatun, R., Akhtar, N. & Rahman, A.S.M.A. 1992. Aquatic weeds as a source of protein in Bangladesh. *Bangladesh Journal of Scientific and Industrial Research*, 27: 103-111.

Marinho-Soriano, E. 2007. Seaweed biofilters: An environmentally friendly solution. *World Aquaculture*, 38(3): 31-33, 71.

Matai, S. & Bagchi, D.K. 1980. Water hyacinth: a plant with prolific bioproductivity and photosynthesis. pp. 144-148. *In* A. Gnanam, S. Krishnaswamy & J.S. Kahn, eds. *Proceedings of the International Symposium on Biological Applications of Solar Energy, 1-5 December 1978, Madurai, India.* Madras, Macmillan.

Matanjun, P., Mohamed, S., Mustapha, N.M., & Muhammad, K. 2009. Nutrient content of tropical edible seaweeds, *Eucheuma cottonii, Caulerpa lentillifera* and *Sargassum polycystum. Journal of Applied Phycology*, 21: 75-80.

McIntosh, D., King, C. & Fitzsimmons, K. 2003. Tilapia for biological control of Giant *Salvinia*. *Journal of Aquatic Plant Management*, 41: 28-31.

McLay, C.L. 1976. The effect of pH on the population growth of three species of duckweed: *Spirodela oligorrhiza, Lemna minor* and *Wolffia arrhiza. Freshwater Biology*, 6: 125-136.

Meske, C. & Pfeffer, E. 1978. Growth experiment with carp and grass carp. *Arch. Hydrobiol. Beih.*, 11: 98-107.

Micha, J.C., Antoine, T., Wery, P. & Van Hove, C. 1988. Growth, ingestion capacity, comparative appetency and biochemical composition of *Oreochromis niloticus* and *Tilapia rendalli* fed with *Azolla*. pp. 347-355. *In* R.S.V. Pullin, T. Bhukaswan, K. Tonguthai & J.L. Maclean, eds. *The Second International Symposium on Tilapia in Aquaculture*, Manila, ICLARM Conference Proceedings.

Mishra, B.K., Sahu, A.K. & Pani, K.C. 1988. Recycling of the aquatic weed, water hyacinth, and animal wastes in the rearing of Indian major carps. *Aquaculture*, 68: 59-64.

Mitchell, D.S. 1976. The growth and management of *Eichhornia crassipes* and *Salvinia* spp. in their native environment and in alien situations. pp. 167-176. *In* C.K. Varshney & J. Rzoska, eds. *Aquatic weeds in Southeast Asia*. The Hague, Junk.

Mitzner, L. 1978. Evaluation of biological control of nuisance aquatic vegetation by grass carp. *Transactions of the American Fisheries Society*, 107: 135-145.

Morioka, K., Naeshiro, K., Fujiwara, T. & Itoh, Y. 2008. Estimation of meat quality of cultured yellowtail *Seriola quinqueradiata* fed *Porphyra* supplemented diet. p. 507. *In Abstracts of World Aquaculture 2008, 19-23 May 2008, Busan, Korea*. Baton Rouge, World Aquaculture Society.

Msuya, F.E. & Neori, A. 2002. *Ulva reticulata* and *Gracilaria crassa*: macroalgae that can biofilter effluent from tidal fishponds in Tanzania. *Western Indian Ocean Journal of Marine Science*, 1: 117-126.

Muller-Feuga, A. 2004. Microalgae for aquaculture. The current global situation and future trends. pp. 352-364. *In* A. Richmond, ed. *Handbook of Microalgal Culture*. Oxford, Blackwell.

Murthy, H.S. & Devaraj, K.V. 1990. Effect of *Eichhornia* based feed on the growth of carps. pp. 9-11. *In* M. Mohan Joseph, ed. *The Second Indian Fisheries Forum Proceedings*, Mangalore, Asian Fisheries Society Indian Branch.

Murthy, H.S. & Devaraj, K.V. 1991a. Comparison of growth of carps fed on *Salvinia* based feed and conventional feed. *Fishery Technology*, 28: 106-110.

Murthy, H.S. & Devaraj, K.V. 1991b. Utility of pistia (*Pistia stratiotes*) in the diet of carps. *Journal of Aquaculture in the Tropics*, 6: 9-14.

Mustafa, M.G. & Nakagawa, H. 1995. A review: dietary benefits of algae as an additive in fish feed. *Israeli Journal of Aquaculture - Bamidgeh*, 47: 155-162.

Mustafa, M.G., Umino, T. & Nakagawa, H. 1994. The effects of *Spirulina* feeding on muscle protein deposition in red sea bream, *Pagrus major. Journal of Applied Ichthyology*, 10: 141-145.

Mustafa, M.G., Umino, T., Miyake, H. & Nakagawa, H. 1994a. Effects of *Spirulina* sp. meal as feed additive on lipid accumulation in red sea bream. *Suisanzoshoku*, 42: 363-369.

Mustafa, M.G., Takeda, T., Umino, T., Wakamatsu, S. & Nakagawa, H. 1994b. Effects of *Ascophyllum* and *Spirulina* meal as feed additives on growth performance and feed utilization of red sea bream, *Pagrus major. Journal of the Faculty of Applied Biological Science, Hiroshima University*, 33: 125-132.

Mustafa, M.G., Wakamatsu, S., Takeda, T., Umino, T. & Nakagawa, H. 1995. Effects of algae meal as feed additive on growth, feed efficiency, and body composition in red sea bream. *Fisheries Science*, 61: 25-28.

Muztar, A.J., Slinger, S.J. & Burton, J.H. 1978. Chemical composition of aquatic macrophytes. II. Amino acid composition of the protein and non-protein fractions. *Canadian Journal of Plant Science*, 58: 843-849.

Nakagawa, H. 1985. Usefulness of *Chlorella*-extract for improvement of the physiological condition of cultured ayu, *Plecoglossus alltivelis* (Pisces). *Tethys*, 11: 328-334.

Nakagawa, H. & Kasahara, S. 1986. Effect of *Ulva* meal supplement to diet on the lipid metabolism of red sea bream. *Bulletin of the Japanese Society Scientific Fisheries*, 52: 1887-1893.

Nakagawa, H., Kasahara, S., Sugiyama, T. & Wada, I. 1984. Usefulness of *Ulva*-meal as feed supplementary in cultured black sea bream. *Suisanzoshoku*, 32: 20-27.

Nakagawa, H., Kumai, H., Nakamura, M., Nanba, K. & Kasahara, S. 1986. Preventive effect of kelp meal supplement on nutritional disease due to sardine-feeding in cultured yellowtail *Seriola quinqueradiata* (Pisces). *Proceedings of the 3rd Symposium Trace Nutrients Research*, 3: 31-37.

Nakagawa, H., Nematipour, G.R., Yamamoto, M., Sugiyama, T. & Kusaka, K. 1993. Optimum level of *Ulva* meal diet supplement to minimize weight loss during wintering in black sea bream *Acanthopagrus schlegeli* (Bleeker). *Asian Fisheries Science*, 6: 139-148.

Nakazoe, J., Kimura, S., Yokoyama, M. & Iida, H. 1986. Effect of the supplementation of alga or lipids to the diets on the growth and body composition of nibbler *Girella punctata* Grey. *Bulletin of the Tokai Regional Fisheries Research Laboratory*, 120: 43-51.

Nandeesha, M.C., Srikanth, G.K., Keshavanath, P. & Das, S.K. 1991. Protein and fat digestibility of five feed ingredients by an Indian major carp, Catla catla (Ham.). pp. 75-81. *In* S.S. De Silva, ed. *Fish Nutrition Research in Asia*. Special Publication No. 5, Manila, Asian Fisheries Society.

Naskar, K., Banarjee, A.C., Chakraborty, N.M. & Ghosh, A. 1986. Yield of *Wolffia arrhiza* (L.) Horkel ex Wimmer from cement cisterns with different sewage concentrations, and its efficacy as a carp feed. *Aquaculture*, 51: 211-216.

New, M.B. & Csavas, I. 1995. The use of marine resources in aquafeeds. pp. 43-78 *In* H. Reinertsen & H. Haaland, eds. *Sustainable Fish Farming*. Rotterdam, A.A. Balkema.

New, M.B. & Wijkstrom, U.N. 2002. *Use of fishmeal and fish oil in aquafeeds: further thoughts on the fishmeal trap*. FAO Fisheries Circular No. 975. Rome, FAO.

Niamat, R. & Jafri, A.K. 1984. Preliminary observations on the use of water hyacinth (*Eichhornia crassipes*) leaf meal as protein source in fish feeds. *Current Science*, 53: 339-340.

Nikolskij, G.V. & Verigin, B.V. 1966. The basic biological characteristics of white amur and bighead and their acclimatization in the water reservoirs of our country. pp. 30-40. In *Herbivorous Fish*. Moscow, Piscevaja Promyslennost.

Ogburn, D.M. & Ogburn, N.J. 1994. Use of duckweed (*Lemna* sp.) grown in sugarmill effluent for milkfish, *Chanos chanos* Forskal, production. *Aquaculture and Fisheries Management*, 25: 497-503.

Olah, J., Ayyappan, S. & Purushothaman, C.S. 1990. Processing and utilization of fermented water hyacinth, *Eichhornia crassipes* (Mart.) Solms. in carp culture. *Aquaculture Hungarica*, 6: 219-234.

Olivares, E & Colonnello, G. 2000. Salinity gradient in the Manamo River, a dammed distributary of the Orinoco Delta, and its influence on the presence of *Eichhornia crassipes* and *Paspalum repens*. *Interciencia* 25: 242-248.

Olvera-Novoa, M.A., Domínguez-Cen, L.J., Olivera-Castillo, L. & Martínez-Palacios, C.A. 1998. Effect of the use of the microalga *Spirulina maxima* as a fish meal replacement in diets for tilapia, *Oreochromis mossambicus* (Peters), fry. *Aquaculture Research*, 29: 709-715.

Opuszynski, K. 1972. Use of phytophagous fish to control aquatic plants. *Aquaculture*, 1: 61-79.

Oron, G. 1994. Duckweed culture for wastewater renovation and biomass production. *Agricultural Water Management*, 26:27-40.

Pádua, M., Fontoura, P.S.G. & Mathias, A.L. 2004. Chemical composition of *Ulvaria oxysperma* (Kützing) Bliding, *Ulva lactuca* (Linnaeus) and *Ulva fascita* (Delile). *Brazilian Archives of Biology and Technology*, 47: 49-55.

Pandey, V.N. & Srivastava, A.K. 1991a. Yield and nutritional quality of leaf protein concentrate from *Eleocharis dulcis* (Burm. f.) Hensch. *Aquatic Botany*, 41: 369-374.

Pandey, V.N. & Srivastava, A.K. 1991b. Yield and quality of leaf protein concentrates from *Monochoria hastata* L. Solms. *Aquatic Botany*, 40, 295-299.

Pantastico, J.B., Baldia, J.P. & Reyes, D. 1985. Acceptability of five species of freshwater algae to tilapia (*Oreochromis niloticus*) fry. pp. 136-144. *In* C.Y. Cho, C.B. Cowey & T. Watanabe, eds. *Fish Nutrition Research in Asia: Methodological Approaches to Research and Development*. Ottawa, IDRC.

Pantastico, J.B., Baldia, S.F. & Reyes, D.M. 1986. Tilapia (*T. nilotica*) and Azolla (*A. pinnata*) cage farming in Laguna Lake. *Fisheries Research Journal of the Philippines*, 11: 21-28.

Peñaflorida, V.D. & Golez, N.V. 1996. Use of seaweed meals from *Kappaphycus alvarezii* and *Gracilaria heteroclada* as binders in diets for juvenile shrimp *Penaeus monodon*. *Aquaculture*, 143: 393-401.

Peters, G.A., Mayne, B.C., Ray, T.B. & Toia, R.E. 1979. Physiology and biochemistry of the *Azolla-Anabaena* symbiosis. pp. 325-344. *In Nitrogen and Rice*. Los Baños, Laguna, Phillipines, International Rice Research Institute.

Pine, R.T., Anderson, L.W.J. & Hung, S.S.O. 1989. Effects of static versus flowing water on aquatic plant preferences of triploid grass carp. *Transactions of the American Fisheries Society*, 118: 336-344.

Pine, R.T., Anderson, L.W.J. & Hung, S.S.O. 1990. Control of aquatic plants in static and flowing water by yearling triploid grass carp. *Journal of Aquatic Plant Management*, 28: 36-40.

Porath, D. & Koton, A. 1977. Enhancement of protein production in fish ponds with duckweed (*Lemnaceae*). *Israeli Journal of Botany*, 26: 51.

Porath, D. & Pollock, J. 1982. Ammonia stripping by duckweed and its feasibility in circulating aquaculture. *Aquatic Botany*, 13: 125-131.

Porath, D., Hepher, B. & Koton, A. 1979. Duckweed as an aquatic crop: evaluation of clones for aquaculture. *Aquatic Botany*, 7: 273-278.

Prabhavathy, G. & Sreenivasan, A. 1977. Cultural prospects of Chinese carps in Tamil Nadu. *Proceedings of the Indo-Pacific Fisheries Council*, 17: 354-362.

Primavera, J.H. & Gacutan, R.Q. 1989. Preliminary results of feeding aquatic macrophytes to *Penaeus monodon* juveniles. *Aquaculture*, 80: 189-193.

Prowse, G.A. 1971. Experimental criteria for studying grass carp feeding in relation to weed control. *Progressive Fish Culturist*, 33: 128-131.

Pullin, R.S.V. & Almazan, G. 1983. *Azolla* as a fish food. *ICLARM Newsletter*, January 1983: 6-7.

Rafiqul, I.M., Jalal, K.C.A. & Alam, M.Z. 2005. Environmental factors for optimisation of *Spirulina* biomass in laboratory culture. *Biotechnology*, 4: 19-22.

Rakocy, J.E. & Allison, R. 1981. Evaluation of a closed recirculating system for the culture of tilapia and aquatic macrophytes. *Proceedings of the Bio-engineering Symposium for Fish Culture*. Bethesda, Maryland, 1: 296-307 Bethesda, Fish Culture Section of the American Fisheries Society and the Northeast Society of Conservation Engineers.

Ramesh, M.R., Shankar, K.M., Mohan, C.V. & Varghese, T.J. 1999. Comparison of three plant substrates for enhancing carp growth through bacterial biofilm. *Aquaculture Engineering*, 19: 119-131.

Ray, A.K. & Das, I. 1994. Apparent digestibility of some aquatic macrophytes in rohu, *Labeo rohita* (Ham.) fingerlings. *Journal of Aquaculture in the Tropics*, 9: 335-342.

Redding, T.A. & Midlen, A,B. 1992. *Fish Production in Irrigation Canals: A Review*. FAO Technical Paper No. 317, FAO, Rome, Italy, 114 pp.

Reddy, K.R. & DeBusk, W.F. 1984. Growth characteristics of aquatic macrophytes cultured in nutrient-enriched water: I. Water hyacinth, water lettuce, and pennywort. *Economic Botany*, 38: 229-239.

Reddy, K.R. & DeBusk, W.F. 1985. Growth characteristics of aquatic macrophytes cultured in nutrient-enriched water: 2. Azolla, duckweed, and Salvinia. *Economic Botany*, 39: 200-208.

Reddy, P.V.G.K., Ayyappan, S., Thampy, D.M. & Krishna, G. 2005. *Textbook of Fish Genetics and Biotechnology*. New Delhi, Indian Council of Agricultural Research. 218 pp.

Rejmánková, E. 1975. Comparison of *Lemna gibba* and *Lemna minor* from the production viewpoint. *Aquatic Botany*, 1: 423-427.

Rejmánková, E. 1979. *The function of duckweeds in fish pond ecosystem*. Trebon, Czechoslovakia, Department of Hydrobotany. 166 pp. (PhD thesis)

Rejmánková, E. 1981. On the production ecology of duckweeds. *Paper presented at the International Workshop on Aquatic Macrophytes, Illmitz, Austria, 3-10 May, 1981.* (unpublished)

Riechert, C. & Trede, R. 1977. Preliminary experiments on utilization of water hyacinth by grass carp. *Weed Research*, 17: 357-360.

Rifai, S.A. 1979. The use of aquatic plants as feed for *Tilapia nilotica* in floating cages. pp. 61-64. *In Proceedings of the SEAFDEC/IDRC International Workshop on Pen and Cage Culture of Fish, 11-22 February 1979, Tigbauan, Iloilo,* Philippines. Iloilo, SEAFDEC.

Robinette, H.R., Brunson, M.W. & Day, E.J. 1980. Use of duckweed in diets of channel catfish. pp. 108-114. *In Proceedings of the 13th Annual Conference of the Southeastern Association of Fish and Wildlife Agencies (SEAFWA)*.

Ruskin, F.R. and Shipley, D.W. 1976. *Making aquatic weeds useful: some perspectives for developing countries*. Washington, D.C., National Academy of Sciences.

Rusoff, L.L., Blakeney, E.W. & Culley, D.D. 1980. Duckweeds (*Lemnaceae* family): a potential source of protein and amino acids. *Journal of Agricultural and Food Chemistry*, 28: 848-850.

Saeed, M.O. & Ziebell, C.D. 1986. Effects of dietary nonpreferred aquatic plants on the growth of redbelly tilapia (*Tilapia zillii*). *Progressive Fish Culturist*, 48: 110-112.

Sahu, A.K., Sahoo, S.K. & Giri, S.S. 2002. Efficacy of water hyacinth compost in nursery ponds for larval rearing of Indian major carp, *Labeo rohita*. *Bioresource Technology*, 85: 309-311.

Said, M.Z., Culley, D.D., Standifer, L.C., Epps, E.A., Myers, R.W. & Bonney, S.A. 1979. Effect of harvest rate, waste loading & stocking density on the yield of duckweeds. *Proceedings of the World Mariculture Society*, 10: 769-780.

Saint-Paul, U., Werder, U. & Teixeira, A.S. 1981. Use of water hyacinth in feeding trials with matrincha (*Brycon* sp.). *Journal of Aquatic Plant Management*, 19: 18-22.

Santiago, C.B., Aldaba, M.B., Reyes, O.S. & Laron, M.A. 1988. Response of Nile tilapia (*Oreochromis niloticus*) fry to diets containing *Azolla* meal. pp. 377-382. *In* R.S.V. Pullin, T. Bhukaswan, K. Tonguthai & J.L. Maclean, eds. *The Second International Symposium on Tilapia in Aquaculture*. Manila, ICLARM Conference Proceedings 15.

Satoh, K-I., Nakagawa, H. & Kasahara, S. 1987. Effect of *Ulva* meal supplementation on disease resistance of red sea bream. *Nippon Suisan Gakkaishi*, 53: 1115-1120.

Schwartz, D.P. & Maughan, O.E. 1984. The feeding preferences of *Tilapia aurea* (Steindachner) for five aquatic plants. *Proceedings of the Oklahoma Academy of Science*, 64: 14-16.

Scott, B. & Orr, L.D. 1970. Estimating the amount of aquatic weed consumed by grass carp. *Progressive Fish Culturist*, 32: 51-54.

Sculthorpe, C.D. 1971. *The Biology of Aquatic Vascular Plants*. London, Edward Arnold. 610 pp.

Senanayake, F.R. 1981. The athko kutu (brush-park) fishery of Sri Lanka. *ICLARM Newsletter*, 4(4):20-21.

Shanmugasundaram, V.S & Balusamy, M. 1993. Rice-fish-azolla: a sustainable farming system. *NAGA, The ICLARM Quarterly*, 16 (2-3): 23.

Shanmugasundaram, V.S. & Ravi, K. 1992. Rice-fish-azolla integration. *NAGA, The ICLARM Quarterly*, 15 (2): 29.

Sharma, K.P. 1981. Solar energy utilization efficiency of Typha wetland. *Current Science*, 23: 1033.

Sherief, P.M. & James, T. 1994. Nutritive value of the water fern *Azolla* for fish. *Fishing Chimes*, 14: 14.

Shetty, H.P.C. & Nandeesha, M.C. 1988. An overview of carp nutrition research in India. pp. 96-116. *In* S.S. De Silva, ed. *Fish Nutrition Research in Asia*, Singapore, Heinemann Asia.

Shireman, J.V. & Maceina, M.J. 1981. The utilization of grass carp, *Ctenopharyngodon idella* for hydrilla control in Lake Baldwin, Florida. *Journal of Fish Biology*, 19: 629-636.

Shireman, J.V., Colle, D.E. & Rottmann, R.W. 1977. Intensive culture of grass carp, *Ctenopharyngodon idella* in circular tanks. *Journal of Fish Biology*, 11: 267-272.

Shireman, J.V., Colle, D.E. & Rottmann, R.W. 1978. Growth of grass carp fed natural and prepared diets under intensive culture. *Journal of Fish Biology*, 12: 457-464.

Singh, S.B., Sukumaran, K.K., Chakrabarti. P.C. & Bagchi, M.M. 1967. Observations on the efficacy of grass carp, *Ctenopharyngodon idella* (Val.) in controlling and utilizing aquatic weeds in ponds in India. *Proceedings of the Indo-Pacific Fisheries Council*, 12: 220-235.

Sipauba-Tavares, L.H. & Braga, F.M.S. 2007. The feeding activity of *Colossoma macropomum* larvae (tambaqui) in fish ponds with water hyacinth (*Eichhornia crassipes*) fertilizer. *Brazilian Journal of Biology*, 67: 459-466.

Skillicorn, P., Spira, W. & Journey, W. 1993. *Duckweed aquaculture: a new aquatic farming system for developing countries.* Washington, D.C., The World Bank. 76 pp.

Somsueb, P. 1995. Aquafeeds and feeding strategies in Thailand. pp. 365-385. *In* M.B. New, A.G.J. Tacon & I. Csavas, eds. *Farm-made Aquafeeds.* FAO Fisheries Technical Paper No. 343. Rome, FAO.

Stephenson, M., Turner, G., Pope, P., Colt, J., Knight, A. & Tchobanoglous, G. 1980. *The use and potential of aquatic species for wastewater treatment. Appendix A: The environmental requirements of aquatic plants.* Sacramento, California State Water Resources Control Board. 655 pp.

Sutton, D.L. 1974. Utilization of hydrilla by the white amur. *Hyacinth Control Journal*, 12: 66-70.

Sutton, D.L. 1990. Growth of *Sagittaria subulata* and interaction with *Hydrilla*. *Journal of Aquatic Plant Management*, 28: 20-22.

Sutton, D.L. & Ornes, W.H. 1975. Phosphorus removal from static sewage effluent using duckweed. *Journal of Environmental Quality*, 4: 367-370.

Tacon, A.G.J. 1987. *The Nutrition and Feeding of Farmed Fish and Shrimp- A Training Manual. 2. Nutrient Sources and Composition.* Brasilia, FAO Field Document, Project GCP/RLA/075/ITA, Field Document 5/E. 129 pp.

Tacon, A.G.J. 1994. *Feed ingredients for carnivorous fish species alternatives to fishmeal and other fishery resources.* FAO Fisheries Circular No. 881. Rome, FAO.

Tacon, A.G.J. 2004. Use of fishmeal and fish oil in aquaculture: a global perspective. *Aquatic Resources Culture and Development*, 1: 3–14.

Tacon, A.G.J., Hasan, M.R. & Subasinghe, R.P. 2006. Use of fishery resources as feed inputs for aquaculture development: trends and policy implications. FAO Fisheries Circular, No. 1018. Rome, FAO. 99 pp.

Tacon, A.G.J. and Metain, M. 2008. Global overview on the use of fish meal and fish oil in industrially compounded aquafeeds: trends and future prospects. Aquaculture (*in press*).

Tacon, A.G.J., Rausin, N., Kadari, M. & Cornelis, P. 1990. The food and feeding of marine finfish in floating net cages at the National Seafarming Development Centre, Lampung, Indonesia: rabbitfish, *Siganus canaliculatus* (Park). *Aquaculture and Fisheries Management*. 21: 375-390.

Tantikitti, C., Rittibhonbhun, N., Chaiyakum, K. & Tansakul, R. 1988. Economics of tilapia pen culture using various feeds in Thale Noi, Songkhla Lake, Thailand. pp. 569-574. *In* R.S.V. Pullin, T. Bhukaswan, K. Tonguthai & J.L. Maclean, eds. *The Second International Symposium on Tilapia in Aquaculture*. Manila, ICLARM Conference Proceedings 15.

Thuyet, T.Q. & Tuan, D.T. 1973. Azolla: a green compost. *Agricultural Problems*, 4: 119-127.

Tuan, N.A., Thuy, N.Q., Tam, B.M. & Ut, V.V. 1994. Use of water hyacinth (*Eichhornia crassipes*) as supplementary feed for nursing fish in Vietnam. pp. 101-106. *In* S.S. De Silva, ed. *Fish Nutrition Research in Asia*, Special Publication No. 9. Manila, Asian Fisheries Society.

Turan, G. 2009. Potential role of seaweed culture in integrated multitrophic aquaculture (IMTA) systems for sustainable marine aquaculture in Turkey. *Aquaculture Europe*, 34(1):5-15.

Valente, L.M.P., Gouveia, A., Rema, P., Matos, J., Gomes, E.F. & Pinto, I.S. 2006. Evaluation of three seaweeds *Gracilaria bursa-pastoris*, *Ulva rigida* and *Gracilaria cornea* as dietary ingredients in European sea bass (*Dicentrarchus labrax*) juveniles. *Aquaculture*, 252: 85-91.

Van der Does, J. & Klink, F.J. 1991. Excessive growth of Lemnaceae and Azolla in ditches observed by false colour teledetection. *Verhandlungen des Internationalen Vereins für Limnologie*, 24: 2683-2688.

Van Dyke, J.M. & Sutton, D.L. 1977. Digestion of duckweed (*Lemna* spp.) by the grass carp (*Ctenopharyngodon idella*). *Journal of Fish Biology*, 11:273-278.

Van Dyke, J.M., Lestie, A.J. & Nall, L.E. 1984. The effects of the grass carp on the aquatic macrophytes of four Florida lakes. *Journal of Aquatic Plant Management*, 22: 87-95.

Van Hove, C. 1989. Azolla *and its Multipurpose Uses with Emphasis on Africa*. Rome, FAO. 53 pp.

Van Hove, C., Baillonville, T.D.W., Diana, H.F., Godard, P., Mai Kodomi, Y. & Sanginga, N. 1987. *Azolla* collection and selection. pp. 77-87. *In Azolla Utilization*. Los Baños, Laguna. International Rice Research Institute.

Varshney, C.K. & Singh, K.P. 1976. A survey of aquatic weed problem in India. pp. 31-41. *In* C.K. Varshney & J. Rzóska, eds. *Aquatic Weeds in South East Asia*. The Hague, Junk.

Venkatesh, B. & Shetty, H.P.C. 1978a. Studies on the growth rate of grass carp, *Ctenopharyngodon idella* (Val.) fed on two aquatic weeds and terrestrial grass. *Aquaculture*, 13: 45-53.

Venkatesh, B. & Shetty, H.P.C. 1978b. Nutritive value of two aquatic weeds and a terrestrial grass as feed for grass carp, *Ctenopharyngodon idella* (Val.). *Mysore Journal of Agricultural Science*, 12: 597-600.

Vroon, R. & Weller, B. 1995. Treatment of domestic wastewater in a combined UASB-reactor duckweed pond system. Wageningen, Landbouwuniversiteit. 110 pp. (Doktoral verslagen serie nr.)

Wagner, G.M. 1997. *Azolla*: A review of its biology and utilization. *The Botanical Review*, 63: 1-26.

Wahab, M.A. & Kibria, M.G. 1994. Katha and Kua fisheries unusual fishing methods in Bangladesh. *Aquaculture News*, 18: 24.

Wahab, M.A., Azim, M.E., Ali, M.H., Beveridge, M.C.M. & Khan, S. 1999. The potential of periphyton-based culture of native major carp kalbaush, *Labeo calbasu* (Hamilton). *Aquaculture Research*, 30: 409-419.

Watanabe, I., Espinas, C.R., Berja, N.S. & Alimagno, B.V. 1977. Utilization of the *Azolla Anabaena* complex as a nitrogen fertilizer for rice. *IRRI Research Paper Series*, 11: 1-15.

Wee, K.L. 1991. Use of non-conventional feedstuffs of plant origin as fish feeds- Is it practical and economically feasible. pp. 13-32. In S.S. De Silva (editor), *Fish Nutrition Research in Asia, Special Publication No. 5*. Manila, Asian Fisheries Society.

Welcomme, R.L. 1972. An evaluation of acadja method of fishing as practiced in the coastal lagoons of Dahomey (West Africa). *Journal of Fish Biology*, 4: 39-55.

Westlake, D.F. 1963. Comparisons of plant productivity. *Biological Reviews*, 38: 385–425.

Westlake, D.F. 1966. Some basic data for investigations of the productivity of aquatic macrophytes. pp. 229-248. *In* C.R. Goldman, ed. *Primary Productivity in Aquatic Environments*. Berkeley, University of California Press.

Wiley, M.J., Pescitelli, S.M. & Wike, L.D. 1986. The relationship between feeding preference and consumption rates in grass carp and grass carp X bighead carp hybrids. *Journal of Fish Biology*, 29: 507-514.

Wilson, J.R., Holst, N. & Rees, M. 2005. Determinants and patterns of population growth in water hyacinth. *Aquatic Botany*, 81: 51-67.

Wilson, J.R., Rees, M., Holst, N., Thomas, M.B. & Hill, G. 2001. Water hyacinth population dynamics. *In* M.H. Hill, T.D. Centre & D. Jianqing, eds. *Biological and Integrated Control of Water Hyacinth*, Eichhornia crassipes. Canberra, ACIAR Proceedings 102.

Wolverton, B.C. 1979. Engineering design data for small vascular aquatic plant wastewater treatment systems. *In Proceedings of the EPA Seminar on Aquaculture Systems for wastewater treatment*. Washington, D.C., EPA 430/9-80-006.

Wolverton, B.C. & McDonald, R.C. 1976. Don't waste waterweeds. *New Scientist*, 12th August, 1976: 318-320.

Xianghua, L. 1988. Research on fish nutrition in China. pp. 92-95. *In* S.S. De Silva, ed. *Fish Nutrition Research in Asia*. Singapore, Heinemann Asia.

Yi, Y.H. & Chang, Y.J. 1994. Physiological effects of seamustard supplement diet on the growth and body composition of young rockfish, *Sebastes schlegeli*. *Bulletin of the Korean Fisheries Society*, 27: 69-82.

Yone, Y., Furuichi, M. & Urano, K. 1986a. Effects of dietary wakame *Undaria penatifida* and *Ascophyllum nodosum* supplements on growth, feed efficiency, and proximate compositions of liver and muscle of red sea bream. *Bulletin of the Japanese Society of Scientific Fisheries*, 52: 1465-1465.

Yone, Y., Furuichi, M. & Urano, K. 1986b. Effects of wakame *Undaria penatifida* and *Ascophyllum nodosum* on absorption of dietary nutrients, and blood sugar and plasma free amino-N levels of red sea bream. *Nippon Suisan Gakkaishi*, 52: 1817-1819.

Zaher, M., Begum, N.N., Hoq, M.E. & Bhuiyan, A.K.A. 1995. Suitability of duckweed, *Lemna minor* as an ingredient in the feed of tilapia, *Oreochromis niloticus*. *Bangladesh Journal of Zoology*, 23: 7-12.

Zhou, Y., Yang, H., Hu, H., Liu, Y., Mao, Y., Zhou, H., Xu, X. & Zhang, F. 2006. Bioremediation potential of the macroalga *Gracilaria lemaneiformis* (Rhodophyta) integrated into fed fish culture in coastal waters of north China. *Aquaculture*, 252: 264-276.

Zimmerman, W.J., Watanabe, I., Ventura, T., Payanral, P. & Lumpkin, T.A. 1991. Aspects of the genetic and botanical status of neotropical *Azolla* species. *New Phytologist*, 119: 561-566.

Zirschky, J. & Reed, S.C. 1988. The use of duckweed for wastewater treatment. *Journal of the Water Pollution Control Federation*, 60: 1254-1258.

Zuccarello, G.C., Critchley, A.T., Smith, J., Sieber, V., Lhonneur, G.B. & West, J.A. 2006. Systematics and genetic variation in commercial *Kappaphycus* and *Eucheuma* (Solieriaceae, Rhodophyta). *Journal of Applied Phycology*, 18: 643-651.

Zutshi, D.P. & Vass, K.K. 1973. Ecology of macrophytic vegetation of Kashmir lake. pp. 141-146. *In* C.R. Varshney and J. Rzòska, eds. *Aquatic Weeds in S.E. Asia*. The Hague, Junk.

Annex 1
Essential amino acid (EAA) composition of aquatic macrophytes

The EAA of some aquatic macrophytes is provided in Table 1. Further information on the EAA composition of *Azolla* and on duckweed is contained in Table 2 and Table 3 respectively.

TABLE 1
Essential amino acid composition of some aquatic macrophytes

Aquatic macrophytes	CP (percent)	EAA (percent of protein)											References
		Arg	Hist	Iso	Leu	Lys	Met	Phen	Thr	Tryp	Val	Tyr	
Alligator weed (*Alternanthera philoxeroides*)	15.1	2.10*	1.10*	1.50*	1.90*	1.50*	0.60*	Trace*	1.60*	-	1.80*	-	Tacon (1987)
Arrowhead (*Sagittaria spp.*)	18.2	1.10*	0.60*	0.90*	1.70*	1.60*	0.20*	Trace*	1.00*	-	1.40*	-	Tacon (1987)
Azolla (*Azolla filiculoides*)	n.s.	6.62	2.31	5.38	9.05	6.45	1.88	5.64	4.70	2.01	6.75	4.10	Buckingham et al. (1978)
Azolla (*Azolla pinnata*) Bangkok strain	n.s.	11.14	2.19	3.64	7.10	5.77	1.27	4.61	2.82	0.23	4.62	-	Almazan et al. (1986)
Canadian pondweed (*Elodea canadensis*), Canada	14.1	6.95	1.35	4.26	7.45	5.68	1.63	4.47	3.76	1.70	5.32	3.48	Muztar, Slinger & Burton (1978)
Curlyleaf pondweed (*Potamogeton crispus*), Canada	12.9	6.36	1.40	4.89	8.14	5.12	2.72	4.81	3.72	0.31	5.74	3.14	Muztar, Slinger & Burton (1978)
Chara sp., Canada	6.1	3.94	0.82	3.28	5.57	3.77	0.82	3.44	3.61	1.48	4.43	2.13	Muztar, Slinger & Burton (1978)
Duck weed (*Lemna minor*), Canada	20.0	5.30	1.60	4.75	8.50	5.65	1.50	4.40	4.40	1.15	5.80	2.85	Muztar, Slinger & Burton (1978)
Eelgrass (*Vallisneria americana*), Canada	18.3	4.26	0.99	4.10	7.16	2.19	1.26	4.92	3.33	1.15	4.70	3.17	Muztar, Slinger & Burton (1978)
Eurasian water milfoil (*Myriophyllum spicatum*), Canada	12.8	7.04	1.87	5.76	9.92	7.37	2.12	7.54	4.75	0.60	7.37	3.90	Muztar, Slinger & Burton (1978)
Oxygen weed (*Hydrilla verticillata*)	n.s.	4.18	1.43	3.89	7.16	4.12	1.63	4.61	3.78	-	4.69	3.55	Boyd (1969)
Sago pondweed (*Potamogeton pectinatus*), Canada	19.7	4.32	1.12	3.55	5.99	6.45	1.02	4.57	3.15	0.92	4.42	2.54	Muztar, Slinger & Burton (1978)
Water hyacinth (*Eichhornia crassipes*)	n.s.	4.55	1.62	3.86	6.78	4.68	1.37	4.09	3.78	-	4.49	2.93	Boyd (1968)
Water lettuce (*Pistia stratiotes*)	n.s.	3.63	1.69	3.99	7.06	5.27	1.35	4.45	3.84	-	4.82	3.19	Boyd (1968)
Water spinach (*Ipomoea aquatica*)	n.s.	3.31	2.66	3.42	6.55	4.56	1.53	5.67	3.92	-	5.27	4.14	Peñaflorida (1989)
Water willow (*Justicia americana*)	17.6	3.00*	1.10*	2.50*	4.30*	2.80*	0.90*	2.80*	2.30*	-	2.90*	-	Tacon (1987)

Arg = Arginine; Hist = Histidine; Iso = Isoleucine; Leu = Leucine; Lys = Lysine; Met = Methionine; Phen = Phenylalanine; Thr = Threonine; Tryp = Tryptophan; Val = Valine; Tyr = Tyrosine)
n.s. = not stated.
*Values expressed as percent DM basis.

TABLE 2
Essential amino acid composition of Azolla species

Azolla species	Amino acids[1] (percent DM)								
	Arg	Hist	Iso	Leu	Lys	Met + Cys	Phen + Tyr	Thr	Val
A. microphylla[2]	1.90	0.47	1.07	2.29	1.62	0.43	2.17	1.13	1.07
A. caroliniana[2]	1.58	0.40	0.85	1.96	1.34	0.46	1.93	1.03	0.86
A. filiculoides[2]	1.04	0.28	0.57	1.42	1.04	0.47	1.29	0.68	0.79
A. nilotica[2]	1.56	0.37	0.84	1.71	1.27	0.52	1.51	0.91	0.81
A. pinnata var. imbricata[2]	1.43	0.33	0.76	1.79	1.15	0.21	1.57	0.86	0.88
A. mexicana[2]	1.33	0.32	0.75	1.66	1.06	0.51	1.45	0.85	0.75
A. pinnata var. pinnata[2]	1.32	0.32	0.81	1.71	0.96	0.23	1.45	0.84	0.97
A. pinnata[3]	1.15	n.s.	0.93	1.65	0.98	0.52	1.69	0.87	1.18

[1] Arg = Arginine; Hist = Histidine; Iso = Isoleucine; Leu = Leucine; Lys = Lysine; Met = Methionine; Phen = Phenylalanine; Thr = Threonine; Val = Valine; Cys = Cysteine; Tyr = Tyrosine); crude protein levels not stated
[2] modified from Cagauan and Pullin (1991)
[3] Alalade and Iyayi (2006)

TABLE 3
Mean essential amino acid values (g/100 g protein) of four species of duckweed[1] compared to FAO reference EAA pattern

Amino acids	Mean ± SD	FAO reference protein
Arginine	4.54 ± 0.64	-
Histidine	1.78 ± 0.42	-
Isoleucine	3.61 ± 0.37	4.2
Leucine	6.68 ± 0.58	4.8
Lysine	4.01 ± 0.43	4.2
Methionine	0.90 ± 0.15	2.2
Phenylalanine	4.16 ± 0.39	2.8
Threonine	3.12 ± 0.40	2.8
Tryptophan[2]	-	1.4
Valine	4.39 ± 0.64	4.2
Tyrosine	2.82 ± 0.44	-

[1] L. gibba, S. polyrrhiza, S. punctata and W. columbiana
[2] Destroyed during analysis
Source: modified from Culley et al. (1981)

Annex 2
Periphyton

Rich periphyton communities boost fish production. The distribution of periphytic fauna shows differences with regard to quantum and seasonal succession. Periphyton-supported aquaculture systems offer the possibility of increasing both primary production and food availability for fish; especially those low in the food chain. The culture of milkfish (*Chanos chanos*), a very popular cultured species in Indonesia, Philippines and Taiwan Province of China, is mainly based on periphytic *"lab lab"* as food, the production of which is enhanced by organic and inorganic fertilization (Juliano, 1985). The *"acadjas"* of West Africa (Welcomme, 1972), the brush parks of Sri Lanka (Senanayake, 1981) and the *"Katha"* fisheries of Bangladesh and India (Wahab and Kibria, 1994) are well-known examples of periphyton-based aquaculture systems.

Dempster, Beveridge and Baird (1993) have reported that Nile tilapia graze more efficiently on periphyton substrates than on micro-particles in the water column. Algal biomass is also higher in periphyton systems. Bhaumik *et al.* (2005) have reported that richness of periphytic structure in closed wetlands results in higher fish production (1 570 kg/ha/year) compared to open system (384 kg/ha/year). Lagoons provided with substrates for periphyton, supports eight times higher algal biomass compared to the surrounding lagoons (Konan-Brou and Guiral, 1994).

A range of substrate-supported aquaculture systems (Table 1) have been developed to reduce the cost of feeding fish (Azim *et al.*, 2002a, 2002b; Keshavanath *et al.*, 2002; Garg, 2005). In these systems additional substrates are provided for the growth of periphyton, which has positive effects on fish production. The association of microorganisms, algae and planktonic organisms attached as periphyton serve as food for fish and also act as an *in situ* water purifier ensuring better living conditions. Wahab *et al.* (1999) have reported 1.8 times higher production of carp kalbaush (*Labeo calbasu*) in ponds provided with scrap bamboo as substrate than from ponds without substrate. Similar results were also observed with rohu (*Labeo rohita*) (Azim *et al.*, 2001), Mahseer (*Tor khurdee*) (Keshavanath *et al.*, 2001) and milkfish (*Chanos chanos*) (Jana *et al.*, 2006). Fish yield is linearly correlated with substrate area (Azim *et al.*, 2004). Garg (2005) has reported that grey mullet (*Mugil cephalus*), milkfish (*Chanos chanos*), pearlspot (*Etroplus suratensis*) and Nile tilapia (*O. niloticus*) are suitable species for periphyton-based brackish water culture systems. Survival and growth of these four fish were higher in substrate-supported periphyton-based culture systems compared to the systems without substrate. The provision of additional substrates in fish culture ponds reduce the use of artificial feed, especially those species that thrive low in the food web.

TABLE 1
Various substrates used in periphyton-based culture system

Fish species	Culture system	Substrate used	Reference
Tilapia	Monoculture	Dense masses of branches	Welcomme (1972)
Sarotherodon melanotheron	Monoculture	Bamboo poles	Hem and Avit (1994)
Labeo calbasu	Monoculture	Scrap bamboo	Wahab *et al.* (1999)
Labeo fimbriatus	Polyculture	Bamboo, jutesticks	Azim *et al.* (2002a)
Labeo rohita	Monoculture	Sugarcane bagasse	Ramesh *et al.* (1999)
Cyprinus carpio	Monoculture	Paddy straw (*Eichhornea* sp.)	Ramesh *et al.* (1999)
Tor khudree	Monoculture	Bamboo poles	Keshavanath *et al.* (2002)